Catastrophic Disaster Planning and Response

Catastrophic Disaster Planning and Response

Clifford E. Oliver
CEM, CBCP

CRC Press
Taylor & Francis Group
Boca Raton London New York

CRC Press is an imprint of the
Taylor & Francis Group, an **informa** business

Front Cover photo credits: Top Left: U.S. Air Force photo by Tech. Sgt. James L. Harper Jr. Top Right and Middle Left: U.S. Navy photos by Petty Officer 2nd Class Justin Stumberg. Middle Right and Bottom Right: DoD photos by Fred W. Baker III. Bottom Left: U.S. Navy photo by Petty Officer 2nd Class Chelsea Kennedy.

CRC Press
Taylor & Francis Group
6000 Broken Sound Parkway NW, Suite 300
Boca Raton, FL 33487-2742

Printed in the United States of America on acid-free paper
10 9 8 7 6 5 4 3 2 1

International Standard Book Number: 978-1-4398-4169-3 (Hardback)

Library of Congress Cataloging-in-Publication Data

Oliver, Clifford.
Catastrophic disaster planning and response / Clifford Oliver.
p. cm.
Includes bibliographical references and index.
ISBN 978-1-4398-4169-3 (hardback)
1. Emergency management. 2. Emergency management--United States. 3. Preparedness. 4. Disasters. I. Title.

HV551.2.O45 2010
363.34--dc22 2010009989

Visit the Taylor & Francis Web site at
http://www.taylorandfrancis.com

and the CRC Press Web site at
http://www.crcpress.com

Dedication

This book is dedicated to my father, Robert Henry Oliver, who passed away in January 2009. As an experienced engineer and scientist, my father instilled in me a compelling desire to be both factual and concise in my thinking and communication. We would often discuss scientific issues of the day and he would test my position, asking if my knowledge was reliable.

Reliable knowledge is knowledge that has a high probability of being true because its veracity has been justified by a reliable, often scientific, method. Reliable knowledge is sometimes called *justified true belief*, to distinguish reliable knowledge from belief that is false and unjustified or even true, but unjustified. As a result of our talks, I must credit my father with helping me to become a stronger, effective applier of the proper scientific method and to become a critical thinker.

Contents

SECTION II ETHICAL, POLITICAL, AND LEGAL ISSUES

Preface

I joined academia late in life. What drew me to it was a long-standing desire to share with others, especially the post-9/11 generation of emergency managers and homeland security specialists, my experiences and lessons learned from my over 25 years of experience in both emergency management and homeland security. My credentials for writing this book come more from my professional knowledge and life experiences than from my academic credentials. The path I took to writing this book is lined with colleagues and friends who believed in me and who saw potential in me that I had yet to fully realize and only recently have come to truly appreciate.

Towards the end of 2008, I was appointed to the federal Senior Executive Service (SES) as part of my continuing career with the Department of Homeland Security's Federal Emergency Management Agency (FEMA). As an emergency management professional, I had spent the better part of twenty years working for FEMA in a number of capacities that allowed me, whether it was Hurricanes Andrew, Katrina, Rita, Gustav, or Ike, the Great Mississippi Flood of 1993, the bombing of the Alfred P. Murrah Federal Building, or the attacks of September 11, 2001, to stand witness to some of the largest disasters and defining moments this nation has faced in recent times. As I joined the SES, I was humbled by the recognition by my FEMA's leadership of my body of professional work and achievement, my contribution to FEMA's success, and my efforts in helping our nation recover from disaster after disaster. Years earlier, I had been recognized as a professional emergency manager by being certified as both a *Certified Emergency Manager* (CEM®) by the International Association of Emergency Managers and a *Certified Business Continuity Professional* (CBCP) by the Disaster Recovery Institute International. These certifications recognized my technical and accomplishment within the emergency management professional, while my appointment to the SES recognized my overall professional accomplishments and my ability to be an effective executive, manager, and leader.

As word spread among my friends and colleagues of my SES appointment, a long-time friend suggested I consider entering the field of collegial instruction. This friend, William W. Sondervan, Ph.D., professor and executive director, Criminal Justice Administration and Affiliate Programs, University of Maryland University College (UMUC), believed that my professional experience, including instructional experience (though limited at the time), made me a good candidate to become an instructor at UMUC and so, in February 2009, I was appointed as an adjunct associate professor in the Emergency Management, Fire Science, and Homeland Security Programs.

As I was new to collegiate academia, I waited patiently for my first teaching assignment. In May 2009, at an academic meeting, Stephen S. Carter, academic director, Emergency Management, Fire Science, and Homeland Security Programs, and I were discussing when I might be assigned a class to teach. I also mentioned that I was "game" for just about any academic assignment. Within weeks, Steve contacted me and asked if I was interested in developing a senior undergraduate course on catastrophic planning and response for the emergency management program. The course was scheduled to be offered in the Fall 2009 semester, leaving me just three months to develop a 400-level course, a process that normally takes six months. Never one to shy away from a challenge, I accepted this assignment. As I raced to put the course together, my literature search found there was no existing comprehensive textbook on the subject of planning and responding to catastrophes. What I did find in my literature search would prove, in the end, to be my ultimate savior in developing this course in those three months. What I found was a course under development by the FEMA Emergency Management Institute's (EMI) Higher Education Program entitled *Catastrophe Readiness and Response*.

While the course materials were still under development, B. Wayne Blanchard, Ph.D., CEM, EMI Higher Education Program manager, was generous enough to share the draft modules with me so I could make use of them in developing what became an equivalent of 300 pages of instructional material for the UMUC course. I am equally grateful to the lead developer of this course, Rick Bissell, Ph.D., professor and Graduate Program director, Department of Emergency Health Services, University of Maryland, Baltimore County, who provided much assistance and exhibited great patience with this neophyte collegial course developer.

One of the greatest challenges in developing the course for UMUC was the lack of a course textbook. As a result, I made use of selective readings from a number of books and a large 300-page commentary was developed. This UMUC course made its debut in the fall of 2009 with me as the instructor. While the students rated the course highly, the lack of the main textbook was noted as a challenge by all involved.

In June 2009, while I was developing the course, I attended the EMI Higher Education Conference and struck up a conversation with Mark Listewnik, from CRC Press, about the lack of a textbook on catastrophic planning and response. It was Mark who first asked why I didn't write such a textbook. To that point, I had not considered myself as part of the solution to this problem. While Mark's proposition was simple, it was a momentous moment for me. I pondered, me, write a textbook? Who am I to be writing a collegial level textbook? Were my professional and personal experiences sufficient enough to form the basis of a high quality textbook?

The more I thought about it, the more the idea of writing a textbook appealed to me. That fall, I explored the possibility of authoring a textbook with Stephen Carter. Having observed my efforts in producing the UMUC course and seeing the impact of not having a main textbook to support delivery of the course, Steve threw his support behind the idea and was instrumental and securing permission from UMUC to incorporate certain aspects of the course material into a future textbook. With support from Steve, Mark and most importantly, my family, late in 2009, I entered into a contract with CRC Press to author this book.

In addition to acknowledging the contributions of those named above, I would be greatly remiss if I failed to acknowledge those who contributed material to the EMI Higher Education Program course entitled *Catastrophe Readiness and Response* as this course is an important source of material for this book. These subject matter experts (in alphabetical order) are:

- Timothy Beatley, Ph.D., Teresa Heinz Professor of Sustainable Communities, Department of Urban and Environmental Planning, School of Architecture, University of Virginia
- Rick Bissell, Ph.D., professor and Graduate Program director, Department of Emergency Health Services, University of Maryland Baltimore County
- Drew Bumbak, MS, Senior Emergency Preparedness /COOP/COG analyst, Science Applications International Corporation (SAIC)
- Thomas Kirsch, M.D., M.P.H., associate professor, director of Operations, Department of Emergency Medicine, deputy director of the Office of Critical Event Preparedness and Response, co-director of the Center for Refugee and Disaster Response, Johns Hopkins University School of Medicine
- Brian J Maguire, Dr.PH, MSA, EMT-P, clinical associate professor, Department of Emergency Health Services, University of Maryland Baltimore County
- David A. McEntire, Ph.D., associate dean, College of Public Affairs and Community Service; associate professor, Emergency Administration and Planning, Department of Public Administration, University of North Texas

- Anthony Oliver-Smith, Ph.D., professor emeritus, Department of Anthropology, University of Florida
- John C. Pine, Ph.D., director of the Research Institute for Environment, Energy and Economics and professor in the Department of Geography & Planning, Appalachian State University
- Jasmin Ruback, Ph.D., president of Ruback Associates, Inc.
- Anna Schwab, JD/MRP, program manager, Center for the Study of Natural Hazards and Disasters, University of North Carolina at Chapel Hill
- Kevin M. Simmons, Ph.D., professor of Economics, Austin College
- Gavin Smith, Ph.D., research professor, executive director, Center for the Study of Natural Hazards and Disasters; executive director, Department of Homeland Security's Center of Natural Disasters, Coastal Infrastructure and Emergency Management
- Myra M. Socher, BS, EMT/P, adjunct assistant professor of Health Care Sciences, The George Washington University School of Medicine and Health Sciences, and adjoint assistant professor of Nursing (Disaster Management), Vanderbilt University School of Nursing
- Tricia Wachtendorf, Ph.D., associate director, Disaster Research Center, University of Delaware; assistant professor, Department of Sociology and Criminal Justice
- Scott Wells, senior Disaster Response consultant

What you will find when you read this book is a practitioner's perspective on planning and responding to catastrophes. That's not to say that there is not a significant level of theoretical and scholarly discourse, but this is tempered by the fact that reality places real-world constraints on our ability to effectively manage in a catastrophe. Lastly, I have made every attempt to produce a book that is up-to-date, but one must realize that the field of catastrophic planning and response continues to evolve both domestically and internationally.

As if this point needed to be proven, as I was putting the finishing touches on the book manuscript, the horrific Haitian earthquake struck. While I did include some preliminary information on the first month of the response, there is no doubt that significant research and investigation will be undertaken in the coming months and years that will dissect how the international community responded to this latest catastrophe. I assume this future research will be one of the bases for a future revision to this book.

I hope you enjoy reading this book as much as I enjoyed authoring it.

Clifford Oliver, CEM, CMCP

Introduction

TURQUOISE BLUE WATERS: THE DECEIVING CALM

While writing this book in the winter of 2009–2010, my family and I took a Caribbean cruise. As we plied the turquoise waters of the Caribbean on our way to the island of Saint Thomas in the U.S. Virgin Islands (USVI), I took some time to reflect on the many months I spent in the USVI in 1995 as part of the Federal Emergency Management Agency's (FEMA) response to Hurricane Marilyn. The island was devastated by the effects of this category 3 hurricane. I got to see, first hand, many of the response, recovery, and reconstruction efforts that result from a catastrophe such as this one. The negative effects of these impacts were amplified greatly by logistical challenges brought on by the impacted area being an island.

I remembered back to the day I arrived on Saint Thomas on a chartered aircraft, within a couple of days after the storm passed. From the moment we landed, the lack of infrastructure was apparent when the copilot informed us that we would need to unload our own baggage from the cargo hold because there weren't any airport personnel available and all the baggage handling equipment had been damaged. As I got off the plane, I scanned the scene and noticed a heap of twisted metal girders and realized that this once was the airport terminal, one of the island's lifelines to mainland USA. We loaded our bags into a caravan of damaged, but operable, vehicles and we drove from the airport to a badly damaged hotel in the small town of Red Hook that offered to house us. Along the way, it was easy to see that many of the island's homes were substantially damaged and the island's critical infrastructure was badly damaged and nonfunctional. Water, sewer, electrical, and phone service all were out and remained out, in some parts of the island, for more than six months. Basic construction materials needed to make rudimentary repairs were nonexistent.

The hurricane had a devastating, cascading effect on the island's economy. With the airport infrastructure nearly destroyed, commercial flights were suspended for more than a month while temporary repairs were made and the Federal Aviation Administration (FAA) and federal security authorities recertified the airport for commercial service. With little-to-no working infrastructure, the physical damage to the hotels and the airport, tourists were unable to visit the island for many months. As a result of the expensive damage, cruise ships diverted to other ports of call as well, further depriving the island economy of critical tourist dollars.

With the loss of infrastructure and the all-important tourists, many on the island became jobless. With the devastation, the sense of hopelessness, and idle time waiting for the situation to improve, the social fabric began to rapidly degrade and looting broke out. The situation degraded to the point where local law enforcement was accused of either being complicit in the looting or at least turning a blind eye to the situation. The U.S. Department of Justice, using authorities almost never used, deployed federal and local law enforcement, under a mutual aid agreement with the federal government, to Saint Thomas to quell the unrest. These outside law enforcement personnel stayed on the island for months until the economy stabilized and workers returned to their jobs.

My time spent working at the Joint Field Office dealing with the innumerable effects brought on by the loss of housing, economic opportunities, and critical infrastructure, left an indelible mark that, to this day, shades my view of how to effectively respond to a catastrophe. In my almost 20 years of federal service, I have been witness to effects of many other large-scale disasters including Hurricane Andrew in South Florida, the attacks of September 11, 2001, and Hurricane Katrina. I have spent countless days working alongside federal, state, and local emergency responders, other government officials, and volunteers, all of us working without a break for weeks, and I observed, first hand, the sad impact as responders reached the end of their mental and physical endurance. In responding to these same events, I have looked into the faces of many disaster survivors and seen the hopeless look I call the *thousand-mile stare.* As an emergency manager, it is my responsibility to help reduce the suffering these people experience.

While those in academia might argue whether these events meet the definition of a catastrophe or are just large-scale disasters, nonetheless, these events have all played an important role in shaping my understanding, view, and opinion concerning how the United States can effectively plan for and respond to a catastrophe.

WHAT THIS BOOK ACCOMPLISHES

This book is designed to fill a gap in emergency management education, namely the issue of events so large and complex that normal disaster preparedness and response strategies, resources, and skills are vastly insufficient. In the United States, academia, emergency management practitioners, and researchers call these events *catastrophes.* This book is intended to be used as a reference for professionals as part of their toolkit, as well as a text for upperdivision undergraduate or master's level courses that serve as an introduction to the field of catastrophe planning and response; it is not and cannot serve as the final resource in a field that is rapidly developing. Once you have utilized this book as part of your professional learning, you are encouraged to stay current on the subject by reading trade and scholarly journals, as well as government reports, by attending industry conferences, and by joining trade organizations, such as the International Association of Emergency Managers and the Disaster Recovery Institute International.

THE CUTTING EDGE

Because catastrophe study and research are relatively new, there are relatively few resources that will be available in the library and in online search engines. For this reason, this book instead relies on several recently published books as well as government reports, scholarly articles, and papers. Most importantly, much of the source material for this book came from the Catastrophe Readiness and Response course that was under development by the Higher Education Program of FEMA's Emergency Management Institute (EMI) (Emmitsburg, Maryland) at the time this book was written. When performing research for the field of catastrophic planning and response, be aware that some of the newer peer-reviewed emergency management journals are beginning to cover the topic of catastrophe planning and response more extensively. These journals may well serve as your most up-to-date source of academic exploration of the topic.

Emergency management has always been a multidisciplinary endeavor. The study of how we prepare for and respond to catastrophes is perhaps even more strongly dependent upon the input of multiple academic and practitioner disciplines if we are going to reach a workable understanding of the issues. While I am a recognized and experienced emergency management professional in the field of emergency management, material incorporated into this book comes from authors from a variety of disciplines and academic backgrounds including emergency medical sciences, social sciences, and, of course, emergency management. I would like to collectively thank these individuals for their contributions to understanding disasters and catastrophic events.

HOW THIS BOOK IS ORGANIZED

Through the five sections of this book, you will be introduced to the many political, legal, and programmatic issues that influence how we plan for and respond to a catastrophe.

- **Section I**: Because it is important to understand the unique characteristics of catastrophes, Section I offers an overview of the history of catastrophic events both in and outside the United States, how catastrophes differ from disasters and emergencies, and how they are all part of the emergency management continuum. We also will explore the varying definitions of catastrophes and their political and societal implications.
- **Section II**: Here you will be introduced to the inherent ethical, political, and legal issues associated with catastrophic events. With respect to ethics, this module addresses the main ethical and value dilemmas and quandaries that one will likely face before, during, and after a catastrophe. We also will explore the legal framework associated with government response to catastrophes, including the use of the military for domestic response, suspension of civil rights, and federal control of industrial output. Political factors, as well as organizational dynamics, are included to provide a basis for understanding the complex environment in which preparing for extreme events may take place. The literature reflected in political science and public administration provides us with insights into conflicts that arise in highly stressful events and the nature of the problems that evolve from our attempts to deal with disasters.
- **Section III**: Here, we will explore the postcatastrophic environment that an emergency manager might experience. We will learn that many, if not all, environs will be altered by the event. Government, industrial, and health and safety processes and systems will cease to operate as intended for extended periods of time. We investigate the impact a catastrophe might have on logistics, critical infrastructure, mass care, and mass evacuation.
- **Section IV**: While we learned in previous parts of the book that a catastrophe will certainly exceed available response, recovery, and reconstruction capabilities, in Section IV, we will explore planning strategies and skills an emergency manager can employ to mitigate the effects of such an event. Issues discussed will include the use of volunteers and unconventional sources of assistance, implications on mass healthcare issues, and an understanding of the political challenges associated with recovery and reconstruction.
- **Section V**: We will learn that, when faced with a catastrophe, an emergency manager will need to utilize crisis leadership skills that he/she may not have had to use before. We will learn how to lead and influence others in a catastrophic situation by increasing our crisis leadership skills and abilities.

WRAP-UP

At the conclusion of this book, you will have a solid understanding of basic catastrophic planning and response, including its historical development, elements, structures, and functions. *Catastrophic Disaster Planning and Response* provides a foundation for the continued study of and critical reflection on catastrophic planning and response.

One final note before we get started: This material can be as difficult to present as it is to study. It is very hard for people to imagine the reality of conditions that are radically different from what they have previously experienced. We know that disasters occur fairly frequently. In recent years, the president has declared about 50 to 60 major disasters in the United States. In more socio-economically developed countries, such as industrialized nations, the vast majority of these events are met with responses that adequately address the needs of those who are affected. These same countries lack experience with catastrophic events that so totally overwhelm societal response mechanisms that victims are essentially left on their own for an extended period of time. We know from history that such events happen and will happen again. For you as professionals, and those of you who are students of the emerging field of emergency management, transitioning your thinking from the disaster context to catastrophes will challenge your ability to alter your paradigms and start thinking in new terms. So, let's get started.

Clifford Oliver

The Author

Clifford Oliver, CEM, CBCP, is a member of the Federal Senior Executive Service and is director of the Acquisition Program and Planning Division in FEMA's Acquisition Management Division. In this capacity, Oliver manages the full acquisition lifecycle of FEMA's most complex and highest priority acquisitions. Prior to occupying this position, he held numerous management and senior technical positions within FEMA, including overseeing the temporary housing program during Hurricanes Katrina and Rita emergency responses. In this capacity, Oliver managed the $5 billion effort that resulted in the installation of 152,000 temporary housing units and the design and construction of over 200 temporary housing group sites scattered along the Gulf Coast from Texas to Alabama. At its peak, there were 10,000 contract personnel in the field installing 1,000 temporary housing units a day, expending approximately $20 million per day. In overseeing this program, Oliver managed the four largest contracts and over 60 of the 100 largest contracts within the Department of Homeland Security (DHS) during fiscal year 2006. Prior to this, he was the director of the Mitigation Division's Risk Assessment Branch where he oversaw such programs as the National Dam Safety and Earthquake Hazards Reduction programs as well as the development of HAZUS (HAZards United States), FEMA's lost estimate model.

Oliver is also an adjunct professor at the University of Maryland University College in the Homeland Security and Emergency Management programs. In this capacity, he is the course chair, author, and lead instructor for the Catastrophic Planning and Response course.

Oliver has over 25 years of engineering, as well as program and emergency management experience, at both the federal and local government levels. He is recognized by the DHS as a Certified Acquisition Professional/Program Manager Level III. In addition, he is recognized by the International Emergency Managers Association as a Certified Emergency Manager (CEM) and by the Disaster Recovery Institute International as a Certified Business Continuity Professional (CBCP). Oliver has BS and MS degrees from the University of Maryland and is currently pursuing his Ph.D.

SECTION I

Introduction to Catastrophic Planning and Response

- Overview
  Because it is important to understand the unique characteristics of catastrophes, Section I will review the history of catastrophic events both in and outside the United States, how catastrophes are different from disasters and emergencies, and how they are all part of the *emergency management continuum*. We will also explore the varying definitions of catastrophes and their political and societal implications.
- Learning Objectives: By the end of Section I, the reader should be able to:
  - Differentiate between major disasters and catastrophes, the planning aspects of each, and their societal impacts.
  - Contrast between different types of catastrophes (e.g., they're all different, but there are commonalities).
  - Understand and interpret the various phases of the emergency+disaster-catastrophe continuum (e.g., emergency → disaster → catastrophe → extinction level event).
  - Contrast between the all-hazards approach and the hazard-unique approach to catastrophe readiness and response.
  - Survey and analyze historical catastrophes and their factors that warrant classification as a catastrophe.

    - Appraise the various aspects of catastrophes and how they can critically affect the U.S. disaster/emergency management system.
    - Compare the theoretical assumptions and policy implications of different definitions of catastrophes.
    - Analyze the impact of conceptions of historical time, culture, and their societal context (including non-United States) on the understanding of catastrophes.
    - Categorize the etiology of typical events in a catastrophe
    - Correlate trends leading toward future events and discuss hypothetical future catastrophic events and their potential effects on modern society, including climate change.
- Outline of Topics
    - Introduction
    - Chapter 1: The Definition of a Catastrophe and the History of Catastrophes
        - Definition of Catastrophe
        - Continuum of *Magnitude*
        - Brief Examples of Historical Catastrophes
        - Potential Catastrophe Effects on U.S. Emergency Management
        - Culture and Definitions of Catastrophe
        - Big Picture
    - Chapter 2: How Catastrophes Differ From Disasters
        - Overview
        - Introduction to Catastrophe Response Planning
        - Past and Future Catastrophes: Their Etiologies and Challenges
        - Potential Future Catastrophes
        - Factors Common in Catastrophes
- Section I Conclusion

CHAPTER 1

The Definition and History of Catastrophes

1.1 LEARNING OBJECTIVES

After reading and studying this chapter, you should be able to:

- Differentiate between major disasters and catastrophes, the planning aspects of each, and their societal impacts.
- Contrast between different catastrophes (e.g., they are all different, but there are commonalities).
- Interpret the emergency–disaster–catastrophe continuum (e.g., emergency → disaster → catastrophe → extinction level event).
- Contrast between the all-hazards approach and the hazards-unique approach to catastrophe readiness and response.
- Analyze historical catastrophes and their factors that warrant classification as a catastrophe.

1.2 KEY TERMS AND PHRASES

- Catastrophe
- Disaster
- Extinction Level Event
- Hypercomplex emergencies
- Quarantelli's six criteria that distinguish catastrophes from disasters
- Continuum of Magnitude
- National Response Framework (Figure 1.1)

1.3 DEFINITION OF CATASTROPHE

BOX 1.1 FEMA DEFINITION OF CATASTROPHE

"... any natural or manmade incident, including terrorism, that results in extraordinary levels of mass casualties, damage, or disruption severely affecting the population, infrastructure, environment, economy, national morale, and/or government functions."

U.S. Department of Homeland Security, National Response Framework. Chapter 2, Response Actions, 42 http://www.fema.gov/pdf/emergency/nrf/nrf-core.pdf

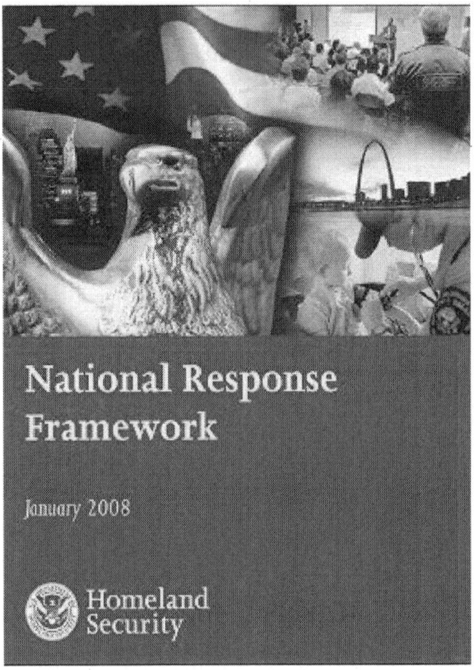

FIGURE 1.1 Cover of the National Response Framework Report.

The first definition presented here is from the Federal Emergency Management Agency (FEMA) and provides the conceptual basis for current FEMA catastrophe readiness activities in several parts of the United States. The inclusion of the term *national morale* could lead to some interesting discussion among practitioners and academia. Just how important is a national psychology in recovering from horrendous events? The attacks of 9/11 may have had a stronger impact on national morale than would have been the case for a natural event with the same number of casualties.

Let's try this definition: "A catastrophe is an event that directly or indirectly affects an entire country, requires national and possibly international response, and threatens the welfare of a substantial number of people for an extended period of time." A synonym used by several European countries is a *hypercomplex emergency*.

This definition brings into discussion the concept that an entire nation is affected for an extended period of time and that international response assistance may be needed. In doing so, it incorporates one of the base concepts of disasters that outside assistance is needed; only this time the "jurisdiction" is much larger. The *hypercomplexity* term is increasingly used in some European countries to describe catastrophes, using a functional perspective of how catastrophes are different from a response viewpoint. Please note that this book will make numerous references to international aspects of catastrophes due to the fact that catastrophes don't respect national boundaries.

Embedded in this definition is the concept that the size of the event is only one of the variables that distinguishes *catastrophes* from *disasters*. Catastrophes differ *in kind* as well as size. By this we mean that their complexity and their various impacts are so significant that the ordinary emergency planning, preparedness, and response tools are no longer sufficient, or may even be counterproductive. One of the core concepts entailed in this definition of *hypercomplexity* is based on the realization that modern social and economic systems are so thoroughly intertwined with multiple diverse actors that no command-and-control system will be effective in bringing all needed resources to bear on a response. Once this is realized, alternative approaches to resource acquisition and utilization can be envisioned.

Famed disaster sociologist E. L. Quarantelli (2006) has developed a list of six criteria that help distinguish catastrophes from disasters. His definition includes indicators that can be recognized at the community level as well as addressing national actors. Quarantelli's definition has become well used in the limited but growing sociological literature on catastrophes.

Let's look at Quarantelli's criteria for a moment. Note that the criteria do not form a definition, but criteria that help the reader see that catastrophes really are different in some observable ways from disasters. Not every criterion needs to be met, and most are qualified by relative terms such as *most*. The first criterion indicates that infrastructure cannot be counted on for local or regional response. However, nearly all writers in this field, both in and out of government, agree that a pandemic could or would qualify as a very serious catastrophe, even

BOX 1.2 QUARANTELLI'S SIX CRITERIA

1. In catastrophes, most or all of a community-built structure is impacted, including facilities of emergency response organizations.
2. Local response personnel are unable to assume normal roles due to losses of personnel and/or facilities and equipment.
3. Help from nearby or even regional communities is not available because all are affected by the same event.
4. Most, if not all, of the everyday community functions are sharply and concurrently interrupted.
5. News coverage is more likely to be provided by national organizations over a longer period of time.
6. National government and very top officials become directly involved.

though it would not touch our physical infrastructure. It would affect our human infrastructure ... those who meet the needs of the population on all levels, thus withdrawing needed services at the very moment they are most needed. In our opinion, this is the real meaning of Quarantelli's first criterion: The resources we most need are directly affected by the catastrophe and rendered unavailable.

The second criterion says the same thing, but focuses on personnel and equipment. The third points out that mutual aid is not to be counted on because the event is of such size that "neighbors" are all similarly affected and unable to come to the aid of others. This is a clear departure from much of the thinking in disaster preparedness and calls for different planning parameters.

The fourth criterion, noting that community functions are sharply curtailed, like the third criterion, indicates that outsiders will be responsible for providing what the affected population needs. This could mean that higher levels of government will begin to take primary responsibility at local levels, or even that nongovernmental organizations (NGOs) from outside of the affected area will become primary service providers. In either case, decisions regarding the use of resources at the local level may be coordinated and decided by outsiders. If we read this right, the combination of criteria two, three, and four breaks with the longstanding concept that outsiders will come in to help in a response that is still directed by capable and knowledgeable locals.

Criteria five and six add more evidence to the shift from local to national or higher levels of participation. The movement of foci away from the local, while still focusing on meeting the needs of local populations throughout the entire affected region, provides a hint at the complexity of catastrophes. This is one of the key concepts of catastrophes.

You should note that none of the definitions we have offered thus far has a single clear tipping point at which an event converts from being a disaster to taking on the characteristics of a catastrophe. This "loose definition" phenomenon is one of the enduring qualities of extreme event preparedness and response.

1.4 CONTINUUM OF *MAGNITUDE*

The graph below provides a simple depiction of the continuum of magnitude ranging from an emergency that can be managed using locally available resources to a so-called extinction level event. You should be able to visualize that the emergencies and disasters, around which most emergency management strategies and tools are designed, are not the upper level of complexity. The catastrophes, around which this book is based, are much less frequent than emergencies and disasters, but represent the highest level of crises for which planning can make a real difference. The extinction level events are beyond effective organized human response, and may be characterized by such events as a major meteorite strike, loss of the Earth's protective ozone layer, or a pandemic with 100% fatality. Without understanding these differences, *catastrophe* could be misunderstood as a synonym for *disaster*. It is important to point out the definitions below are not finite. What might be described as an *emergency* in a large jurisdiction with many resources might be, by definition, a *disaster* in a jurisdiction with limited resources.

BOX 1.3 CONTINUUM OF MAGNITUDE

Emergency Disaster Catastrophe Extinction Level Event

→

1.4.1 Summary Definitions

Emergency: An event, usually sudden, that puts at risk the life or well being of at least one person. Local emergency response resources are adequate to meet the immediate needs of those who are affected by the incident. The response is directed/coordinated by personnel from within the same jurisdiction as the responding agencies.

Disaster: An emergency involving multiple people, of such magnitude that local response resources are *not* adequate to meet the immediate needs of those who are affected by the event, requiring that additional resources be brought in from outside jurisdictions. The response is directed/coordinated by personnel from within the jurisdiction where the event occurred, but many of the responders may be from other jurisdictions, increasing the challenge of response coordination.

Catastrophe: Use one or a combination of the definitions offered above. The response is from so many different jurisdictions, levels of government, and different kinds of organizations, and the needs of the affected population are so diverse and spread out that no single entity can coordinate it all. Many needs will go unmet, at least in the short term.

Extinction Level Event: An event so severe that humans may not survive. No organized useful interventions can be anticipated.

1.4.2 Planning Dilemma: All-Hazards versus Hazards-Unique

U.S. emergency management uses an all-hazards approach to planning and preparedness because:

- Core response management systems are similar for most disaster types.
- It reduces confusion if all responses have the same basic organization.
- It's less expensive.

The difference between disaster and catastrophe planning and response is a bit hard for many to accept, although students beginning their career may have an easier time with it than "journeymen" emergency managers. We start the discussion with a brief overview of the major reasons that FEMA and other government agencies have moved toward an all-hazards approach. It is worth noting that the all-hazards approach is not universally accepted throughout the world as the gold standard, although much of Western Europe and virtually all former British colonies use it. As a sideline, you may be interested in exploring how the British terror threat system works by visiting http://www.mi5.gov.uk.

Students taking a course in planning and responding to a catastrophe should be familiar with the history of the development and adoption of the all-hazards approach in the United States. We intend only to mention the main points in this quick overview.

As a whole, the all-hazards approach to preparing for and managing the response to emergencies and disasters is robust and useful. However,

BOX 1.4 DOWNSIDE OF THE ALL-HAZARDS APPROACH

- Has limited ability to properly prepare for and manage events that require full participation by many private and nongovernmental organizations that are not subject to government authority structures.
- Delegates specific event-type planning to an annex.

its uniform structure across disaster types limits its usefulness in very large or very unique events for which the best response may require different kinds of coordinating and management structures than those that are embodied in the all-hazards approach. For some kinds of hypercomplex events, roles need to be specified in advance regarding what each organization, or type of organization, is going to do. This may be particularly the case when the event type, such as a massive pandemic, is the domain of specialists who normally reside outside of the world of emergency management. The complexities of such a response are so immense as to require a plan that is fully dedicated to that event type, rather than being an annex to a general emergency operations plan.

Numerous researchers have, over the past decade, questioned the wisdom of focusing so much of the U.S. emergency preparedness and response system on an all-hazards command and control model. Drabek and McEntire (2002), Tierney (2002), Wachtendorf (2004), and Waugh (2007) all argue that the current model assumes superior government knowledge and effective reach, and underestimates the need for innovative, emergent, and even spontaneous efforts at local levels in large complex events. It is enough in this introductory chapter to note that well-respected researchers question the command and control-based all-hazards system that is currently the dominant model in the United States.

BOX 1.5 USE OF HAZARD-SPECIFIC APPROACH FOR CATASTROPHES

- Allows greater depth of planning for hypercomplex events so that the planning does not have to be initiated after event onset.
- Allows better focus on some specific event types with peculiar parameters, such as pandemics.
- Hazard-unique planning allows the development of a single plan to serve many jurisdictions.

The advantages of a hazards-unique approach are laid out above. Planning is a way of allowing consequences and decision making to be considered prior to an event's onset. In hypercomplex events, the decisions must address a level of complexity that does not readily lend itself to a small group of response managers who lack expertise in the specifics of the event. More time is needed than is available. Good preevent planning by organizations with expertise relevant to the event type can minimize the time needed to make decisions when the catastrophe is underway. Both the Redlener and Posner books argue persuasively for multijurisdictional, event-specific planning for potential catastrophes. Indeed, perhaps the strongest advantage of the hazards-unique approach to catastrophe response planning is the ability to develop one plan that covers many jurisdictions and levels of government simultaneously. Of course, this works only for catastrophes that can be foreseen.

The National Response Framework (NRF) takes a hazard-specific approach to catastrophic planning and response through hazard-specific annexes as well as the catastrophic incident. We will explore the catastrophic planning response aspects of the NRF further in subsequent modules. To learn more about the NRF Catastrophic Incident Annex, go to: http://www.fema.gov/pdf/emergency/nrf/nrf_CatastrophicIncidentAnnex.pdf.

Later chapters of this book provide in-depth examples of the reasons for changing the planning paradigm for catastrophes to a hazards-unique approach. It is worth noting at this introductory point in this book, however, that FEMA's catastrophe preparedness program in Florida and the New Madrid Seismic Zone (NMSZ) is embracing the multijurisdictional hazards-unique approach and working to fit it into the currently existing all-hazards command and control system.

1.5 BRIEF EXAMPLES OF HISTORICAL CATASTROPHES

1.5.1 1755 Lisbon, Portugal, Earthquake and Tsunami

This section of the chapter is designed to provide some brief examples of catastrophes in history. We hope that this will help the reader to realize that catastrophes are real and affect real people, and to also get a feeling for some of the difficult dynamics that result from catastrophes. The events we've chosen to emphasize here all have numerous accounts written about them; you may choose to familiarize yourself with some of the details. The 1755 combination earthquake and tsunami that hit Lisbon and other areas of Portugal provide an example of an event that changed history for more than just the people who were immediately

BOX 1.6 SITUATION (LISBON)

- 8+ Richter scale earthquake struck Lisbon at 9:40 a.m. on All Saints Day, followed by a massive tsunami some 40 minutes later.
- Fire followed that was uncontrollable.
- Between 60,000 and 100,000 deaths out of a population of 270,000.
- Lisbon and many other coastal communities were destroyed.
- All-important docks and port facilities were lost along much of the coast.

Long-term effects:

- Substantial economic decline occurred for several decades.
- It decimated Portugal's colonial ambitions.
- It greatly exacerbated internal political tensions in Portugal.

affected. This catastrophe occurred before it was common for assistance to arrive from other countries, but the event was well known and written about throughout Europe and colonial North America at the time, and, in fact, impacted some of the philosophical and existentialist writings that are still in vogue today. The postearthquake, decades-long struggle to restore its economy led Portugal to virtually give up its then thriving ambitions to become a major colonial power, thus affecting areas of both Africa and Latin America that may have otherwise come under durable Portuguese colonial power. Portugal was never again an important power within Europe. This event also stimulated the development of purposeful seismic-resistant building design.

1.5.2 Hurricane Mitch: 1998

Hurricane Mitch is an event that occurred within the lifetime of current college students, some of whom may even remember hearing about it. This hurricane would have been a catastrophe in almost any environment; it was a huge storm that dropped over 3 m of rain over a large area. There is virtually no place in the world that could absorb that much rain in three days without substantial damage. The effect was enhanced in Honduras (Figure 1.2), Nicaragua, and El Salvador, all

BOX 1.7 SITUATION (HURRICANE MITCH)

- Category 5 storm "parked" over Honduras for three days in late October of 1998.
- Dropped over 3 m (10 ft) of rain on Honduras.
- Massive floods and landslides killed between 10,000 and 16,000 people. To this day, 8,000 people remain unaccounted for and are presumed dead.
- Most of the country's bridges and many roads were washed out.
- Coastal banana plantations also wiped out.
- Topsoil washed to sea due to the heavy rainfall.
- It took a minimum of seven years to regrow banana trees.
- Farming production is still lagging.
- Infrastructure rebuilding still underway with international help.
- Estimates: 30 years of lost economic development.

FIGURE 1.2 Aerial photos of Tegucigalpa, Honduras after Hurricane Mitch.

developing countries in Central America with uncontrolled deforestation, poorly protected transportation infrastructure, vulnerable power transmission systems, and a landscape consisting of steep mountains and flat agricultural valleys and coastal plains. The net effect was poor drainage of a massive amount of water; numerous landslides; mass loss of bridges, roads, and power lines; and the loss of a huge percent of the all-important cash crop of bananas.

At least 10,000 to 16,000 people lost their lives, and uncounted hundreds of thousands lost their homes and livelihoods. In Honduras, a large percentage of rural and numerous urban health centers were destroyed. At this writing, now over a decade after the storm, many significant components of the Honduran infrastructure have still not been adequately repaired, despite generous amounts of international assistance. The economy is estimated to have been set back some 30 years.

This is an event that meets all of the definitions of catastrophe: None of the affected nations had sufficient resources to respond effectively alone, and all were affected so seriously that mutual aid was impossible. The governments were severely diminished in their effectiveness, and basic life-support systems were rendered ineffective. Transportation, communication, and some power supply systems are still struggling a decade later to reach their previous status, even with significant outside assistance. The suffering will continue for some years yet.

1.5.3 Drought/Famine in India: 1965–1967

BOX 1.8 SITUATION FAMINE

- Monsoon rains essentially failed for three consecutive years.
- Water storage and irrigation systems were insufficient.
- Water tables fell in the first half of the 1900s as a result of British colonial policies favoring deforesting and planting export crops.
- Drought led to food crop failures; an estimated 1.5 million people died despite foreign food assistance.

The first two historical above examples were caused by rapid-onset events. We now turn to a slow-onset catastrophe that had a much higher cost in human lives than either of the two other cited catastrophes. The drought and resulting famine in India from 1965 to 1967 did not result in infrastructure damage (other than temporary damage to croplands), but did result in an uncontrolled loss of human lives and untold suffering. India was incapable of responding effectively; only the significant input of international food assistance kept millions more from perishing. This kind of slow-onset event may be a precursor of threats in the not-too-distant future, as climates change and human populations find themselves ever closer to a margin of existence that can be easily breached. The major point of this example is to emphasize that slow-onset events, while perhaps not as dramatic as earthquakes or mega-storms, can constitute equally or even more devastating catastrophes. We also may want to point out that some of the variables that contributed to the famine were the result of poor land use policies by the British colonial authorities in India. In other words, this event was not solely the result of a weather change, but came about due to decades of environmental destruction at the hands of mankind.

1.6 POTENTIAL CATASTROPHE EFFECTS ON U.S. EMERGENCY MANAGEMENT

Some important observations can be made at this point concerning the potential effects a catastrophe might have on the U.S. emergency management system. Local and regional emergency management personnel may be victims themselves and unable to fulfill roles. Localities may be isolated for an extended period of time without internal capabilities. Therefore, decisions may be made at federal or distant regional levels without local input. Even if local emergency management personnel are available to respond, they may find the complexity of the event beyond their capacity to manage. Loss of normal governing capacity may lead to local or regional chaos. With respect to infrastructure, communications and data transmission and transportation systems may fail. Life-supporting supplies and services may be unavailable locally for an extended period.

1.7 CULTURE AND DEFINITIONS OF CATASTROPHE

The following discussion will take the definitions of catastrophe from the abstract conceptual level to a more concrete understanding of what catastrophes might do to the abilities of emergency managers in the United States and their capacity to coordinate and support an effective response to the events. Culture can influence the way we define catastrophe, and definitions can influence the way we prepare for and respond to catastrophes. The point of this discussion is not to make all seem hopeless, but rather to point out in concrete terms that normal planning and operational assumptions will likely not work in catastrophes. Catastrophes are different in character, not simply just of a larger magnitude of the same dynamics found in disasters. Once you have captured this concept, you will be primed to successfully comprehend the remainder of this book.

We begin with a short discussion concerning culture and definitions of catastrophe primarily to introduce the reality that the definitions we have posited earlier in this chapter are primarily to help scholars, policy makers, and program designers better understand the general parameters of the phenomenon we call "catastrophe." However, in a multicultural nation like the United States, the way that the general population responds to the concept may be heavily influenced by cultural beliefs and assumptions.

Most U.S. emergency managers, policy makers, and politicians come from a background that values the concept of determinism. This leads to the belief that we can make a difference in the outcome of events by injecting certain targeted activities and resources. Without this belief, there would be no emergency management. However, not every culture embraces determinism, which can lead to a "fatalist" approach to philosophy that would devalue investing time and resources into preparedness for events that might not happen in any given place. In the mixture of cultures in the United States, there are a certain number of people who subscribe to a fatalist or passive view of life. To put it another way: "It won't happen here. If it does happen here, it won't happen to me. If it does happen to me, it won't be that bad. If it is that bad, then I couldn't have done anything about it anyway, so why try?"

Another kind of culture that emergency managers have to deal with is organizational culture. This is one that affects emergency managers directly. Among other characteristics is the tendency of bureaucracies to keep resources to themselves, even though sharing resources with other organizations might result in a better outcome for society. The slang term *stovepiping* is one way to describe the difficulties emergency managers face when they attempt to involve many different kinds of organizations in a single-funded project. All too often, the funding stream allows only a single kind of agency to receive funding for a planning or exercise project. This cultural narrowness clearly has implications for catastrophe preparedness and response in which it is recognized right from the beginning that many different kinds of organizations need to be productively involved in the processes of planning, testing, and responding.

One of the major differences between disaster preparedness and catastrophe preparedness is the concept that national-level resources will be sufficient to manage the needs resulting from any disaster. In catastrophe preparedness, it is assumed that assistance may be needed from outside the national borders, even for a wealthy nation like the United States. A culture that perpetuates the sense of national sufficiency and complete independence ignores the true interdependence of modern national and international economies, and stands in the way of addressing the hypercomplexity of catastrophes.

Over the past few decades, there has been frequent mention in the sociology, anthropology, development, and political literature of a "culture of dependency." This is controversial, but is basically characterized by a mindset that assumes that an organization or government will meet the basic needs of people. This might be found in economically disadvantaged people in some parts of the United States, or in the majority of residents of some European countries that provide cradle-to-grave social support services. This carries potential danger for catastrophe response planners who may assume that a population is prepared and capable of fending for itself for a certain period of time before organized resources can reach those who need them.

Indeed, some cultural definitions of *catastrophe* that emphasize hopelessness or inevitability may decrease self-protective action. Recent research reported by Ripley (2008) and others indicate that catastrophe survivors are more likely to be made up of people who believe they can make a difference in the outcome of events by way of their own purposeful actions. It is this belief in the utility of purposeful preparedness and actions that underpins this book.

1.8 THE BIG PICTURE

This discussion is just the beginning of this book, and already there is much to think about. The points made here in the "big picture" list below are not meant so much as a summary of what has already been presented in this opening discussion, but rather a brief overview of some of the challenges. In closing this chapter, you should be left with the sense that catastrophes present a challenge to our society and to emergency managers, one that you need to take personally and for which to make preparations.

BOX 1.9 SUMMARY OF BIG PICTURE ISSUES

- Disasters and catastrophes have been around as long as humans (even longer … previous extinction events).
- Humans in wealthy societies have some protection against catastrophes, but it is far from guaranteed.
- Some cultures have extensive experience with catastrophe, e.g., India, Indonesia, Haiti.
- The current populations in the United States and Europe have little personal experience with catastrophe and may exhibit disbelief that such a thing could befall them.
- Catastrophe responses require such a high level of coordination that preparedness activities must be given a long lead time if there is any chance they will be successful.
- Politicians and citizens are often unwilling to dedicate funding to threats they cannot see in the immediate future.
- The challenge: Prepare for events people don't want to think about, using resources they don't want to dedicate. If the events do happen and we're unprepared, you will be blamed.

1.9 DISCUSSION QUESTIONS

1. Differentiate between the definitions of a catastrophe and a disaster.
2. Why are there varying definitions of a catastrophe?
3. Why is the Continuum of Magnitude important to understanding catastrophes?
4. What kind of issues can arise when an *all-hazards approach* is taken to planning for a catastrophe?
5. From the brief discussion of the history of catastrophes, what kinds commonalities come to mind?

REFERENCES

Drabek, T.E. and D.A. McEntire. 2002. Emergent phenomena and multi-organizational coordination in disasters: Lessons from the research literature. *International Journal of Mass Emergencies and Disasters* 22(2): 197–224.

Posner, R.A. 2004. *Catastrophic risk and response*. New York: Oxford University Press.

Quarantelli, E.L. 2006. Catastrophes are different from disasters: Some implications for crisis planning and managing drawn from Katrina. Social Sciences Research Council Web site: www.ssrc.org (accessed on January 10, 2010).

Redlener, I. 2006 *Americans at risk: Why we are not prepared for mega-disasters and what we can do now*. New York: Knopf.

Ripley, A. 2008, *The unthinkable: Who survives when disaster strikes—and why*. New York: Random House.

Tierney, K.J. 2002. Lessons learned from research on group and organizational responses to disasters. Paper presented at Countering Terrorism: Lessons Learned from Natural and Technological Disasters, Academy of Sciences, Washington, D.C., February 28–March 1.

Wachtendorf, T. 2004. Improvising 9/11: Organizational improvisation in the World Trade Center disaster. PhD diss., University of Delaware.

Waugh, W.L. 2007. Terrorism and disaster. In *Handbook of disaster research,* eds. H. Rodriguez, E.L. Quarantelli, and R.R. Dynes, 388–404. New York: Springer.

CHAPTER 2

How Catastrophes Differ from Disasters

2.1 LEARNING OBJECTIVES

At the end of this chapter, the reader should be able to:

- Appraise the various aspects of catastrophes and how they can critically affect the U.S. disaster/emergency management system.
- Compare the theoretical assumptions and policy implications of different definitions of catastrophes.
- Analyze the impact of conceptions of historical time, culture, and their societal context (including non-United States) on the understanding of catastrophes.
- Categorize the etiology of events in a catastrophe.
- Correlate trends leading toward future events and discuss hypothetical future catastrophic events and their potential effects on modern society, including climate change.

2.2 KEY TERMS AND PHRASES

- Catastrophic Planning
- Emergency Operations Plan
- Etiology of Catastrophes
- National Incident Management Systems (NIMS)

2.3 OVERVIEW

Chapter 2 is designed to bring reality to the conceptual definitions of catastrophe presented in Chapter 1 by way of three topic discussions:

1. A description of many of the ways in which catastrophes and disasters are categorically different from each other.
2. A description of several historical catastrophes.
3. A description of several potential future catastrophes that may strike the United States.

At this point in the book, the objective is to familiarize the reader with the concept of catastrophe in a relatively concrete manner by exploring some past and potential future catastrophes, and looking at some of the commonalities. This is not yet the time to explore the relationship between sociological theory of disaster and catastrophes; this discussion takes place in more detail later on in the book.

A brief note on vocabulary: In Chapter 1, we presented some definitions of catastrophe and contrasted them with disasters and emergencies, moving downward on the intensity/complication scale, and extinction-level events moving upward. This kind of distinction is only a decade or so old in common academic and government/practitioner usage and continues to evolve. Older publications will likely not use the same vocabulary in the same way, potentially leading to confusion if you, the reader, as well as those who are practicing emergency managers are reading multiple texts on this subject. For example, much of what we now call *catastrophes* may appear as *crises, disasters,* or *megadisasters* in older publications. The reader is encouraged to look at the meaning behind the vocabulary that one finds in printed sources and relate that meaning to the terminology we are using in this book. It is clear that today's terminology also may not survive totally intact into future decades. For example, many Europeans' use of the term *hypercomplexity* may find a much wider use in coming years outside Europe. Again, the message to the reader is to look for the basic concepts behind the terminology used and not be too sidetracked by the exact words used.

Section I of this book provides both crucial information and reinforces the concept that catastrophic events have antecedents that can be identified and, in some cases, mitigated. After reading Section I, the reader should come away with the concept that the causal factors leading to many catastrophes can be recognized early enough to plan and prepare for responses that are based on the realization of the hypercomplex character of catastrophes. References for the historical catastrophes described in this chapter are not exhaustive, but will give the readers a good start should they want to significantly deepen their understanding of a given event. Please note that most of the public health-related references were chosen for their combination of accurate material and low level of technical jargon since this book is not focused on public health technical issues.

It is worth noting that most experience with catastrophes has been historical and outside of the United States. For this reason, many of the examples will come from outside of the country, but the principles pertain across borders. The reader should not expect to be knowledgeable about catastrophes as part of their own personal human experience because these events are infrequent. The reader should, therefore, enter into this discussion with an open mind.

BOX 2.1 PRINCIPLES OF DISASTER RESPONSE PLANNING

- Hazard, risk, and vulnerability analysis assumes hazards are in local proximity and can be assessed.
- Resource assessment assumes resources will be available nearby.
- Current planning methods assume jurisdictional authorities and their incident management system will remain viable.
- Agencies generally plan separately.
- Most jurisdictions assume an all-hazards general operations plan, complemented by scenario-based annexes.
- Test and exercise the plans.

This discussion is a brief review of principles and steps that upper division students, who have studied emergency management, should already know. There is no need to belabor the points here, but a quick review is needed in order to juxtapose the following discussion on catastrophe response planning. The reader should take note of the assumptions, particularly the focus on the locality, i.e., the hazards that are planned for are those that exist in or predictably visit the locality, such as hurricanes. It is also worth noting that hazards assessments typically have no longer than a 5- to 10-year outlook. Note that the Federal Emergency Management Agency's (FEMA) Developing and Maintaining State, Territorial, Tribal, and Local Government Emergency Plan (Figure 2.1), which many jurisdictions use as their basic template for developing their Emergency Operations Plan (EOP) and hazards response planning, does not mention in the "hazards unique planning" section the concept of looking into the future for developments, such as climate change or continued poor land

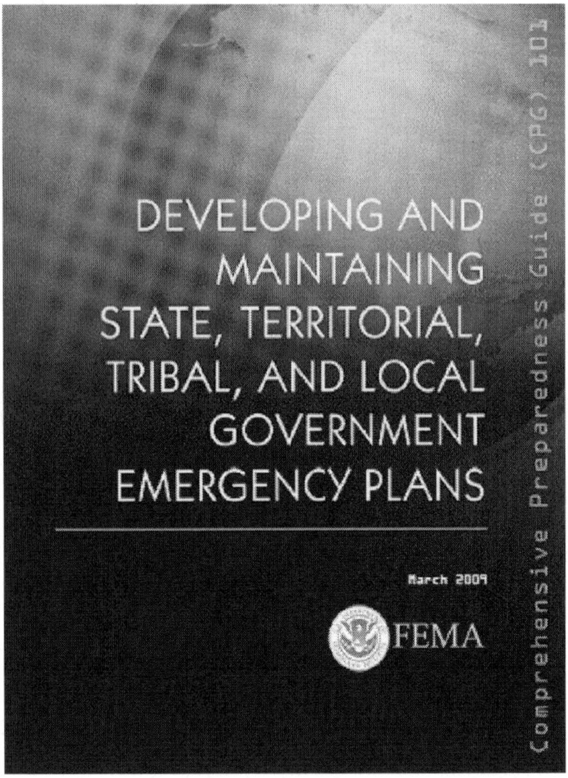

FIGURE 2.1 Cover of FEMA's Developing and Maintaining State, Territorial, Tribal, and Local Government Emergency Plans document. (FEMA).

use planning, that could lead to changes that could significantly increase the risk or intensity of certain hazards. Risk levels are apparently assumed to be static over time and not related to changing climactic conditions (to be discussed later in this commentary). Nor does the guide promote the development of multijurisdictional plans for foreseeable potential catastrophes (see http://www.fema.gov/about/divisions/cpg.shtm).

If the reader is not up to speed, now is a good time to refresh his/her memory as to the steps taken to develop plans for potential catastrophes, and the content included in a typical jurisdictional EOP. For additional review and to support the planning concepts explored in this book, spend a moment to refresh your knowledge of emergency preparedness principles by going to the National Response Framework (NRF) (Chapter 2, p. 27) where there is a discussion on the preparedness cycle with a pictorial model (Figure 2.2).

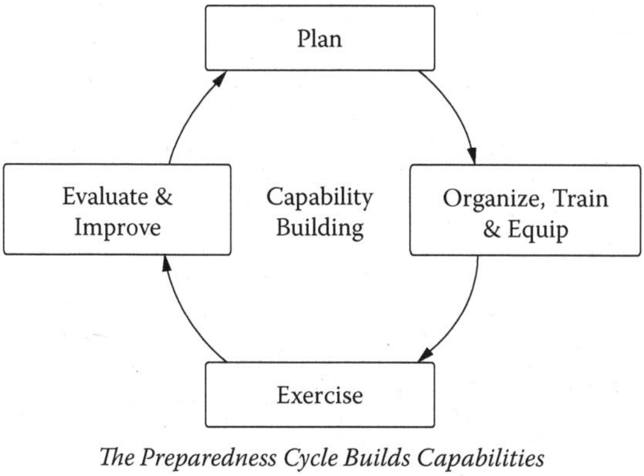

FIGURE 2.2 Preparedness Cycle Building Capabilities from the National Response Framework document. (NRF Figure 2, p. 27.)

2.4 INTRODUCTION TO CATASTROPHE RESPONSE PLANNING

Now we start the process of examining how catastrophe response planning differs from disaster response planning. One of the barriers to considering and planning for catastrophic events is the common mindset that a hazards assessment should focus only on those hazards that are resident in, or particular to, the jurisdiction. Many of the hazards likely to result from global climate change cross many jurisdictions and are not particular to any of them. Pandemics are emblematic of events that know no boundaries.

There are numerous examples of hazards profiles that are changing due to overall global climate change. The point in each of these cases is that current planning methods assume a relatively static hazards profile for a given jurisdiction, and do not encourage planners to either think about changing profiles or multijurisdictional exposure to a given hazard.

Most preparedness work has historically focused on rapid onset events. Many of the historical and future catastrophic events, however, are characterized by a relatively slow onset, even though their actual human toll often supersedes most rapid onset events. Several slow onset events are presented in the upcoming discussion on past and future potential catastrophes in order to drive this point home. One of these potential events is the potential loss of Lake Mead, which straddles the Arizona/Nevada border, and serves as a primary source of water for an estimated 22 million people in four states (*Science Daily*, 2008; Scripps News, 2008). Planning for response to slow onset events presents some advantages logistically, but may require a radically different approach in terms of political and policy strategies.

This discussion makes the point in three different ways that planning and preparedness for catastrophes needs to be both a multijurisdictional and multilevel effort. As noted before, FEMA's Developing and Maintaining State, Territorial, Tribal and Local Government Emergency Plan assumes hazards and events are contained within a given jurisdiction, as if they respected politically drawn boundaries. This assumption is likely true for the vast majority of events that call for first responders or even the activation of an Emergency Operations Center (EOC). However, this is not the case for catastrophes, for which even states as jurisdictions may well be too small to contain the event. This reality renders the current predominant planning methodology inadequate for protecting the needs of the population in catastrophic events.

Two more points that break the mold of routine emergency response planning are:

1. The very definition of catastrophe indicates that national resources are stretched or overwhelmed. In the United States, we do not normally plan for assistance to come in from outside of the country, but catastrophes are of such magnitude that such considerations must be taken into account the planning process. Significant aid was offered from other nations in the aftermath of Hurricane Katrina, but very little of it was allowed by the U.S. government, due mostly to interagency disagreement on what should be allowed and from whom. At the same time, some components of our domestic response to Katrina were clearly insufficient. Foreign offers of assistance could have been handled differently if the Department of Homeland Security, the Department of Health and Human Services, and the Department of State had previously planned a multiagency protocol for accepting and utilizing emergency assistance from beyond our borders.
2. The extreme character of catastrophes, including the very fact that panjurisdictional and even pannational response may be required, means that planning will only be effective if it focuses on the requirements of the specific event type. For example, while planning for response to a pandemic and catastrophic earthquake may have some similarities, the destructive effects and technologies needed to respond are so different from each other that significantly different response, recovery, and reconstruction strategies are needed.

The major point in this discussion is that many of the upcoming probable causes of catastrophes we will soon explore are only effectively understood through the findings of ongoing academic and scientific work. Familiarity with the sciences, therefore, takes on a much more important role than has previously been assumed in emergency management. It also is important to note that likely upcoming catastrophic events cover so many jurisdictions and provoke so many needs that their complexity will require hazard-unique planning and exercising. This is, in itself, complex and potentially expensive. Section IV of this book will describe some of the new multijurisdictional catastrophe response planning and exercise techniques that are being developed and implemented by FEMA for a high-category hurricane in Florida and a New Madrid Seismic Zone (NMSZ) major earthquake in the central United States. While complex, the experience of FEMA and its state and local partners to date indicates that such hazard-specific catastrophic planning is politically appealing as well as both feasible and necessary.

2.5 PAST AND FUTURE CATASTROPHES: THEIR ETIOLOGIES AND CHALLENGES

At this point, we move from abstract concepts to the realities of historical examples of catastrophes as well as likely future threats. We include the word *etiologies* as an attempt to broaden the multidisciplinary vocabulary of the emergency management community. The word *etiology* comes from the fields of public health and medicine, and means the causes of, and typical pathways of a disease or other pathological state. As such, this is a very useful term for emergency managers looking into the future, who can envision calamities as pathologies with distinct causes, pathways, and consequences. In medicine, knowledge of the etiology of a pathology helps bring understanding regarding where successful interventions can be made. Each example presented in this discussion was selected because it:

- Was clearly an overwhelming event for those present.
- Affected many people both directly and indirectly, in many different ways, and over a significant period of time.
- Represents phenomena that could reoccur at any time.

2.5.1 Middle Ages Black Plague

The bubonic plague (Black Death) is an example of a catastrophic event that radically changed European and world history and is thought of by many historians as a seminal event in changing the role of government vis-à-vis the well-being of their citizens. Prior to the bubonic plague,

BOX 2.2 SITUATION: MIDDLE AGES BUBONIC PLAGUE

- A massive change in world trade patterns coupled with an over-populated Europe suffering from 50 years of famine.
- Famine was partially due to rapid climate change toward colder, less predictable weather.
- A new microbe, *Yersinia pestis,* entered Europe at this time via ships from Asia by way of flea-infected shipboard rats.

few western governments (such as they were) saw themselves as responsible for protecting their populous from anything except human invaders. Over the many decades of the Black Death, governments began to engage in both prevention and relief activities, clearly a progenitor of both modern public health and emergency management. For more on this, see Benedictow (2004).

The bubonic plague is a disease that is still around today, caused by the bacterium *Yersinia pestis*, which can infect a variety of mammals. The typical etiology of *Y. pestis* is that it is transmitted from wild rodent hosts (rats, squirrels, etc.) to humans via the bite of fleas that have infested the rodents and picked up the bacterium in their blood meals. Also, *Y. pestis* can be transmitted, although with considerably less frequency, by contact with the flesh, blood, sputum, or pus of infected humans or animals (Heyman, 2008). The disease can kill people in several ways, including massive body-wide sepsis and pneumonia. It is common that victims turn a dark color as the disease progresses, hence the term *Black Death*. The case fatality rate can run from 50% to nearly 100%. Mysteriously, *Y. pestis* can be highly infectious sometimes and not so on other occasions, and, due to it high rate of transmission and death, it has been targeted for development as a potential biological weapon (Orent, 2004).

At the time of the onset of the Black Death, Europe had been undergoing numerous significant changes, several of which teamed with the introduction of *Y. pestis* to wipe out huge portions of many local populations. Successful European agriculture had led to a significant population increase, which, in turn, led to more crowded living conditions. European–Asian commerce had begun to grow. A sudden change to a colder climate decreased crop yields and food availability to the large population, leading to famine at the time that *Y. pestis* entered Europe via shipboard rodents coming from the East. The European population lacked experience with the plague and scientists believe that many in the populace had compromised immune response systems due to hunger.

The result was a massive disease outbreak that lasted decades and eventually resulted in structural changes to society.

The consequences of this catastrophe are virtually unimaginable in today's social consciousness—loss of up to 70% of the population. There wasn't any germ theory-based medicine at the time and no treatments that could offer relief. One of the scenes in Monty Python's *Holy Grail* movie, depicting carts being pulled through city streets to collect the bodies of those who died overnight, is probably pretty accurate (minus the jokes).

Note in this discussion the introduction of mitigation and prevention activities, even though the etiology of the pathogen was not at that time understood. At the end of all the suffering, the Black Death led to several improvements:

- Governments began to take some responsibility for the well-being of their people.
- Some mitigation actions, such as quarantine and social distancing, became known as ways of decreasing disease propagation.
- The huge loss of population, especially among the landless peasant class, led to revised labor relations, giving rise to a "middle" class of merchants and artisans who worked for themselves, not for aristocracy. Given the relative scarcity of labor, workers were able to demand more recompense and better living conditions from their employers.

Note that modern medicine is capable of *decreasing* the lethality of plague, but not stopping it. More worrisome is the specter of the use of genetically altered *Y. pestis* as a bioweapon, against which none of the current control strategies would be efficient.

2.5.2 Little Ice Age in Europe

BOX 2.3 SITUATION: LITTLE ICE AGE IN EUROPE

- From roughly 1300 to 1850, the climate in the Northern Hemisphere became significantly colder, with three minima: 1650, 1770, and 1850.
- Crop-dependent populations hungered and starved.
- Famine decimated the Scandinavian population in Greenland.

The "little ice age" in Europe was a slow onset event of 3.5 centuries' duration. The effects for any particular locality differed over the years and they differed also from region to region, but they were serious. For more information, see Fagan (2000).

Researchers tell us that climate change can bring about local and regional catastrophes. This ice age was relatively mild in terms of the actual temperature change, less than 1° C, according to the Intergovernmental Panel on Climate Change (Intergovernmental Panel, 2001). The calamity it provoked, however, was significant. Humans had been able to greatly expand their population, based largely on increasingly successful agricultural techniques. The downside, then as now, is that we were also very dependent upon continuing success in growing food, which is ultimately largely dependent on temperature, the availability of water, and viable soils. Many of the crops people used from the 1300s through the 1800s were little different from what we depend on today, with a relatively narrow range of temperature and moisture for optimal growth. When crops failed during this period, they did not just fail locally, but across a wide territory, making it difficult or impossible to bring in food from the outside to stem starvation.

Some European populations sealed their own fates by refusing to change their eating habits. For example, Scandinavians, who had begun several colonies on Greenland, starved to death when their crops and grass-fed cattle failed, while the local native people survived eating marine mammals and fish. The losses of some European populations exceeded 50%, such as was the case in Iceland. It was not only in Europe that people succumbed to starvation or related disease. While the records of deaths in North America were poor at that time, there is evidence of some groups, including Native Americans, banding together to avoid starvation (NASA, 2001).

It is worth noting that, while modern society has better heating and food distribution systems, we are generally just as dependent on favorable weather to grow crops as we were in the 1600s. One of the paradoxical potential sequelae of climate change is a rapid onset ice age, which would likely make it quite difficult to support agriculture and other life-support activities affecting 300 million Europeans. For more about this, see Schwartz and Randall (2003).

2.5.3 Irish Potato Famine

This catastrophic event was a combination of natural insult (the potato fungus) and poor decisions made by both the English rulers and the Irish peasants. Prior to the onset of the blight, Ireland had been occupied by England, and virtually all land was forcibly transferred from Irish

BOX 2.4 SITUATION: IRISH POTATO FAMINE

- A combination of the English confiscating Irish farmland, the virtually complete dependence of a huge percentage of the Irish population on potatoes as their base food stock, and the accidental importation of a rapidly spreading fungus (or blight) left vast hunger, starvation, disease, and death in Ireland.
- The blight and crop failure became evident in 1845.
- The British response was confused. First they provided some imported Indian corn, but then stopped and assumed a laissez-faire approach, believing things would sort themselves out.

farmers to new English landlords, essentially making Irish farmers peasants without land rights. The new landlords demanded the planting of wheat and oats crops for export to England, diminishing the land available for Irish agricultural self-sufficiency, thus placing the Irish population in a position of significantly enhanced vulnerability.

Even after the potato blight became a clear reality with obvious consequences for the landless Irish peasants, most of the new British landowners did not allow the Irish to expand the amount of land used for subsistence farming, nor did they allow the Irish to consume the cash crops of grains that were grown on Irish soil for English sale and consumption.

The British government formed an office to provide a response to the Irish potato blight, but then provided this program with very little in terms of staffing or supplies. For a short while, the British provided Indian corn imported from the Americas to the Irish, but then discontinued this practice. While it is fair to say that there were concerned groups of people in England who tried to apply political pressure to force their government to act in a more decisive way, the majority of the British political class reached the conclusion that things in Ireland would sort themselves out on their own.

In this discussion, you see that the potato blight was confounded by British demands and mismanagement, as well as even more increased pressure from nature in the record cold winter. While British troops were guarding the outshipment of grains for English consumption, the Irish were starving. Seeing no alternatives, many Irish started to relocate themselves to other locations by whatever means they could find available.

Here we see that the Irish population made both reasonable and deadly poor decisions in their desperation. Huge numbers of starving peasants left the country, often going to North America or Australia as indentured servants, with a significant percentage of them dying while in transit. Those who remained behind made the fatal mistake of once again planting potatoes as their primary subsistence crop. You may read into this history some conclusions about the willingness or ability of desperate populations to make rational decisions and take helpful actions to enhance their own survival in situations of catastrophic losses. The effects of malnourishment on brain function are well known and may be considered when making decisions regarding response and recovery operations for populations facing starvation.

This is in no way a phenomenon that could be considered an historical relic. If we look at some of the catastrophes that are underway in Sub-Saharan Africa at the present time, we see that the starvation, war, and genocide in places like Darfur, Somalia, Sudan, and the Ethiopia–Eritrea region are a combination of long-term drought, loss of arable land, competition for scarce survival resources, and the use of military and other means of force to assure the predominance of one group over another.

2.5.4 The 1918–1919 Influenza Pandemic

BOX 2.5 SITUATION: THE 1918–1919 INFLUENZA PANDEMIC

- A strain of the Influenza A virus with which humans had no previous experience hit in three waves in 1918 and 1919 with rapid worldwide spread and cumulative mortality that is estimated by various sources to be in the 40 to 100 million range, which represented 3 to 6% of the world population at the time.
- Symptoms started like typical influenza, but often rapidly progressed to severe pneumonia.

The difference between an epidemic and a pandemic is that the epidemic has limited geographical reach, while a pandemic has worldwide spread (even if not all regions are affected equally). Influenza is a viral disease that typically has annual outbreak cycles, with each year's outbreaks generally affecting people who had no previous experience with that particular version of the virus (or one similar to it). The virus that caused the 1918–1919 pandemic was genetically different enough that virtually no one had immunity to it. The disparity in fatality numbers is due to the poor recording of death statistics in many countries at that time, exacerbated by the destruction and disorganization that resulted from World War I.

This discussion of the 1918–1919 pandemic is intended to help you appreciate the potential that a new influenza pandemic that would result in a catastrophic pandemic is just as possible today as was the bubonic plague of the 1300s. Pandemics also present an interesting set of challenges for emergency managers who are contemplating catastrophe preparedness, in that the lead responders to the pandemic would be medical and public health people who know comparatively little about emergency management. However, the public health community, as is described in Section IV of this book, does not have the resources to mount and coordinate a societal response to a pandemic, and will have to rely on emergency managers who know comparatively little about the tools and strategies of epidemiology, public health, and medicine. Coordination in this kind of catastrophe (that attacks humans rather than physical infrastructure) will be so complicated that the only approach to assuring any kind of effectiveness is to engage in serious multidisciplinary, multiagency preplanning and exercising. For more information on this catastrophe, see Kolata (2001).

Researchers believe this pandemic began at a military base in Kansas, having been passed from pigs to humans in the area. The military base provided a densely populated locale for incubation and propagation of the virus, and also a portal for out-migration to other areas of the world as the troops were assigned elsewhere. This particular version of the influenza virus was unusual in its behavior in that it attacked healthy young adults with more frequency and more strength than people in older age groups. Because of military movements at that time, toward the end of World War I, the virus moved quickly to Europe and then on to other parts of the world. After a large outbreak in Spain, Americans took to calling it the Spanish flu, mostly unaware that the disease likely started in the United States. The end of the war stimulated the migration of millions of people, either out of war zones and back to their homes, or out of their home territory toward countries deemed safer, thus providing yet another pathway for the virus to be transported. It is likely that the presence of postwar hunger and poor sanitation also contributed to the seriousness of the disease once it took hold in a new "virgin" population.

In the United States, clinical care facilities were rapidly overwhelmed, as were also mortuaries and morgues. Figure 2.3 illustrates death rates in 1918 as compared to the average death rate for 1911 through 1917. The spike in mortality among 20- to 40-year-olds is quite clear. Alternate care sites were set up in several East Coast cities, but throughout the country many people chose to suffer at home rather than go to a huge ward of the very sick. Many jurisdictions in the country established social-distancing rules and forbad gatherings of people. Some jurisdictions also attempted, some successfully, to isolate themselves entirely from outsiders. In a 2007

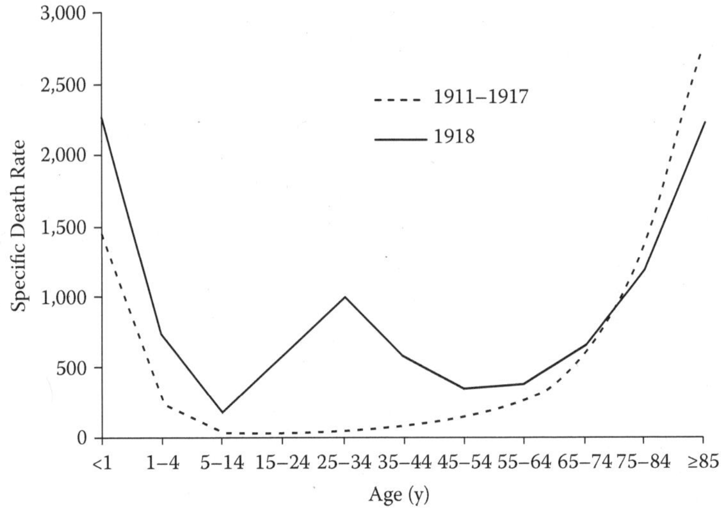

FIGURE 2.3 Taken from Taubenberger, J. and Morens, D. *1918 Influenza: The Mother of All Pandemics*. 2006.

article in the *Journal of the American Medical Association*, Markel et al. (2007) demonstrated that American cities that adopted the practice of using selected isolation and quarantine during the 1918–1919 pandemic significantly decreased transmission of the disease.

Note that some characteristics of life in the United States during that time enabled people to do what might be very difficult or even impossible today, namely, essentially isolate themselves from outsiders during the peaks of the outbreaks. During that time period, food came from nearby and many, if not most people, had stores of food they had preserved from their own gardens. The economy was more organized around local production and consumption. Today's economy is much more globalized, i.e., centralized and interdependent, with widespread use of "just-in-time" delivery models of crucial resources. This makes the entire economy much more vulnerable to disruptions in critical functions, which would be exacerbated by a very large percentage of the population not having a self-sufficient storage of basic foodstuffs. It is not inconceivable that a pandemic could result in widespread hunger and a significant temporary disruption of the monetary economy, in addition to the already horrendous suffering and loss of human life.

Just as has been the case in previous pandemics, the actual death toll of the 1918–1919 influenza pandemic is not precisely known (Figure 2.4). Historians and epidemiologists are comfortable with the assertion that over 500,000 people died of the disease in the United States, but an exact number will never be known. It is more difficult to come to exact

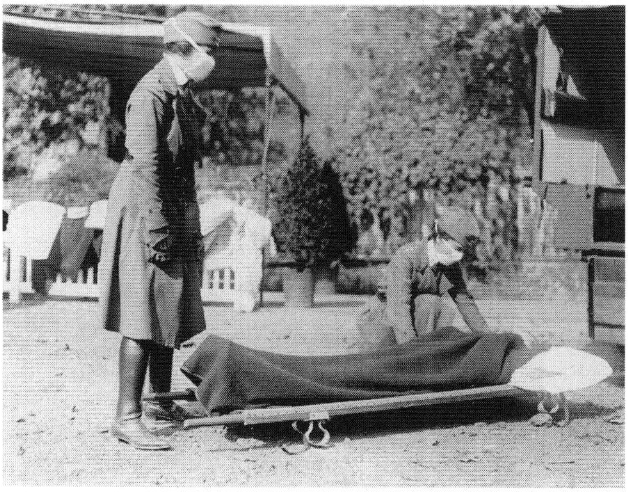

FIGURE 2.4 Demonstration at the Red Cross Emergency Ambulance Station in Washington, D.C., during the influenza pandemic of 1918. (Photo courtesy of the National Photo Company Collection, Library of Congress.)

numbers of those lost in other countries. For many years, historians estimated a figure somewhere around 20 million. However, in more recent studies with more effective information gathering and better epidemiologic models, that figure has been reestimated to fall between 40 and 100 million worldwide (Johnson and Mueller, 2002). Keep in mind that there were less than one-third as many people on the planet then as compared to now.

2.5.5 Tsunami of December 26, 2004

BOX 2.6 SITUATION: TSUNAMI OF DECEMBER 26, 2004

- On December 26, 2004, a 9.1 to 9.3 earthquake off northwestern Sumatra, Indonesia, caused a tsunami with waves as high as 65 ft (20 m) nearest the epicenter.
- At least 200,000 people are believed to have died and 10 million became homeless and displaced.

This 2004 tsunami was one of the deadliest natural disasters in recorded history. Indonesia, Sri Lanka, India, and Thailand were the hardest hit. The waves devastated many areas in the East Indian Ocean basin, particularly the nearby coast of northern Sumatra, the Andaman and Nicobar Islands, and the east and south coasts of Sri Lanka. Areas of southeastern India and southwest Thailand were also hard hit. Deaths and destruction occurred as far away as the coasts of Somalia and Madagascar in Africa, and minor sea level changes were measured as far away as San Diego, California, Iquique, Chile, and Atlantic City, New Jersey. (To view a simulation of the spread of the tsunami wave across the Indian and Pacific Oceans, go to: http://www.noaanews.noaa.gov/video/tsunami-indonesia2004.mov. You will need to have Quicktime® software installed on your computer to watch the animation.)

This tsunami was not the first to cause great losses of life. One of the most destructive tsunamis to occur during historical times followed the explosive eruption of the volcano Krakatoa in the East Indies on August 27, 1883, when over 36,000 people were killed as a result of the wave. Tsunami waves were up to 100 feet in height. Its passage was traced as far away as Panama. It is believed that a 0.6-mile-wide asteroid that struck the ocean southwest of New Zealand about 1500 CE created a tsunami that reached heights of more than 425 feet (*Columbia Encyclopedia*, 2009).

The plight of the many people and countries affected by the tsunami of December 26, 2004, prompted a widespread humanitarian response. In all, the worldwide community donated more than $7 billion in humanitarian aid.

2.5.5.1 Humanitarian Need

The tsunami exacted a heavy toll on coastal communities in the region. In India and Thailand, government and civil society organizations were able to mobilize resources and responded as quickly as possible. India also provided assistance to neighboring countries and was the first nation to respond by sending naval ships and personnel to the neighboring countries due to their proximity. The people and governments in the nations of Sri Lanka and Indonesia were to some extent overwhelmed by the enormity of the catastrophe, especially in inaccessible areas.

The first tasks of the governments and humanitarian aid agencies were to ensure access to food and clean water, and medical care for the injured. The World Health Organization warned that the number of deaths from preventable diseases, such as cholera, diphtheria, dysentery, and typhoid, could rival the death toll from the disaster itself. These diseases are largely spread by loss of normal sanitary facilities, the shared use of inadequate facilities in makeshift refuges, and the lack of clean water.

Many usual sources of water were spoiled by saltwater, broken by the force of the tsunami, or contaminated with bodies of dead people or livestock, requiring water purification equipment or the trucking of portable water into the affected region. Other high priorities were delivery

FIGURE 2.5 U.S. military delivering supplies to December 2004 tsunami victims. (U.S. Department of Defense)

of medical supplies and personnel to overwhelmed hospitals and clinics, tent shelters and clothing to people who have lost their houses and belongings, and food, especially baby food (Figure 2.5). Several governments appealed for body bags to assist in the safe disposal of corpses.

The Tsunami Evaluation Coalition (TEC) has carried out a series of evaluations of the response and published an initial findings report in December 2005 and a final report in July 2006 (TEC, 2006). This report found that while initial needs were broadly met, in part by local actors, there was room for improvement in the way that agencies were meeting ongoing needs. Key areas for improvement in the current agency responses were identified as: (1) their engagement with local actors; (2) transparency, communication with, and accountability to the affected populations; and (3) transparency toward their donors. Despite a number of unique factors, the well-funded tsunami response provides a significant opportunity for the aid community to learn how to improve its performance in future catastrophic responses.

2.5.5.2 Criticism of Tsunami Aid Donors

On December 27, 2004, UN Undersecretary-General for Humanitarian Affairs Jan Egeland reportedly categorized charitable contributions of rich countries as "stingy" (*Washington Times*, 2004), but was widely misinterpreted in the press as categorizing the response to the tsunami in this manner. Speaking at a press conference later, Egeland stated, "It has nothing to do with any particular country or the response to this emergency. We are in early days and the response has so far been overwhelmingly positive" (Reuters News Service, 2004).

In late December 2004, the U.S. government added another $20 million to its original pledge of $15 million, bringing the total up to $35 million, not including direct aid to be rendered by naval vessels dispatched to the region. Initially, the U.S. Navy dispatched P-3C Orion patrol aircraft and an aircraft carrier to assist with relief operations. The P-3C surveillance aircraft conducted survey operations, including search-and-rescue efforts, and cargo planes shuttled supplies from Bangkok to affected areas in the region to shelter the living and dry ice to preserve the dead (Margesson, 2008).

On December 31, 2004, the U.S. pledge was increased tenfold to $350 million, with President Bush saying that that amount will probably increase. He also signed a decree ordering flags to be flown at half-mast during the first week of the New Year.

Serious concern was raised that the international relief effort would falter if nations did not honor their pledges. On January 3, 2005, UN Secretary-General Kofi Annan urged the donor nations to ensure that their pledges will be fully honored, pointing to previous cases where "we got lots of pledges, but we did not receive all the money" (Secretary-General Kofi Annan's statement, 2005).

On January 5, 2005, as countries jockeyed to make large donations, Canadian Radio reported that Jan Egeland said, "I'd rather see competitive compassion than no compassion," adding that too many countries were making pledges that may never arrive. Following the previous year's Bam, Iran, earthquake, which killed 26,000 people, Iranian officials claim to have received just $17.5 million of the $1 billion originally pledged. In mid-March, the Asian Development Bank reported that over $4 billion in aid promised by governments was behind schedule. Sri Lanka criticized the nations and organizations that clamored to pledge donations: "Not a penny had come through yet. We are doing the relief work with our government money. Sri Lanka is still waiting for the money pledged by the donors. Money pledged by the people has been pledged to the NGOs" (BBC News, 2005).

In the same interview, Sri Lanka's Foreign Minister, Laxman Kadirgamar, stated, "A lot of aid that has been coming in lately is, I'm afraid ... I'm sorry to say ... not very useful. For instance there was a container full of teddy bears. They're obviously given with good will, nobody says no to that.... The patience of tsunami-affected nations are being stretched.... Now the government had worked out a scheme that until April 26, 2009, everything that has come, everything that will be on the seas will be admitted tax free. After that, no!" Kadiragamar went on to say, "For instance, we do not need rice, we are expecting a bumper harvest, anyone who sends rice is wasting his time and money."

Many commentators claim excessive and competitive donor responses threaten less dramatic, but equally important, relief efforts elsewhere. "While everyone opens up their coffers for these disasters, the ongoing toll from malaria, AIDS, and tuberculosis is much larger than these one-time events," said Enriqueta Bond, president of the US Burroughs Wellcome Fund. "We would do more good to invest in prevention and good public health measures, such as clean water." Tony Blair, the U.K. prime minister at the time, also expressed concern that tsunami aid could detract from other pressing development needs. He pointed out that there was a disaster comparable to a preventable tsunami every week in Africa, where 10,000 people die daily from AIDS and malaria alone.

2.5.5.3 Criticism of Tsunami Aid-Receiving Countries

In the early stages, before the extent of the disaster was clear, Sri Lanka refused Israel's offers of aid, objecting to the inclusion of 60 Israeli soldiers in the 150-person mission planned by Israel's army, to set up field hospitals, including internal medicine and pediatric clinics, an Israeli army spokesman reported to BBC. The Israeli humanitarian organization later sent a jumbo jet carrying 18 tons of supplies to Sri Lanka, however, and a rescue-and-recovery team from the Jewish ultra orthodox organization, ZAKA, arrived in Sri Lanka with equipment used for identifying bodies as well as supplies of body bags. Corruption, bureaucracy, and nationalism hampered the humanitarian response in Indonesia (*Fox News*, 2005). On January 12, 2005, the Indonesian government put restrictions on the movement of journalists and aid workers, ostensibly for their protection from Acehnese insurgents. However, there were concerns that this was a clumsy attempt by the government to gather control over, and credit for, relief efforts in an attempt to gain a political edge over the rebels.

In Sri Lanka, as of February 10, 2005, only 30% of those eligible for assistance had received any aid and there were allegations of local officials giving aid only to their political supporters, some of whom were not victims of the tsunami. The Sri Lankan government set up a Special Complaint Unit for citizens to record grievances.

2.5.6 Hurricanes Katrina and Rita

BOX 2.7 SITUATION: HURRICANES KATRINA AND RITA

- Two category 5 (at one point) hurricanes hit the U.S. Gulf Coast in 2005.
- The eye of Hurricane Katrina hit some 30 miles east of New Orleans, Louisiana, on August 29, bringing with it a storm surge above 20 feet in many places.
- The morning after Katrina's landfall, several of the levees protecting below-sea level New Orleans failed, flooding the majority of the city and significant portions of surrounding suburbs.
- More than 1,500 people were killed as a result of the storm. Hurricane Rita followed on September 24, making landfall on the Texas-Louisiana border, forcing the evacuation of much of the area around Houston and diverting resources from response to the needs of Katrina victims.

While it is, perhaps, debatable as to whether Hurricane Katrina and its aftermath qualify as a catastrophe or a large-scale disaster, we add it here as a large event with many catastrophe-like characteristics, one that resides in the collective memory of all Americans. It is important to understand the various causes of the response going so badly, including the overwhelming character of some aspects of the events, poor understanding by federal authorities of their own planning documents, poor relationships between levels of government, poor local preparedness, power plays and political interference with incident command at many levels, and misunderstandings among the public as to what government would or could do. At the time of this writing, numerous new research publications on Hurricane Katrina and its aftermath continue to become available through the Disaster Research Center at the University of Delaware and the Natural Hazards Center at the University of Colorado, among others. You are encouraged to peruse the titles of available papers and review those that speak to the issues most important to you. While many of these papers were not written with the concept of catastrophe in mind, many others were, at least as part of the background conceptual model in which the research was conducted.

One of the characteristics of catastrophes is their multijurisdictional coverage. Hurricane Katrina damaged transportation and communications routes in four states, which greatly complicated the tasks of damage assessments, needs assessments, logistics planning and coordination, and communication from those who were directly affected to those who could provide assistance. Coastal economies in the impacted states were significantly impacted, from gambling/tourist interested along the Mississippi coast to the fishing industry in all four states to the petroleum and chemical industries spread out along the Gulf coast.

Healthcare became an issue almost immediately, both for the lack of availability and the need to safely evacuate patients out of threatened (preevent) or damaged (postevent) hospitals and nursing homes to new locations. The damage was so widespread that solutions were not to be found in an adjacent county or parish. In the end, many patients were transported as far away as the Midwest, East Coast, and Southwest regions of the country, often without medical records or a responsible companion or family member.

It is important to note that, in New Orleans, local decisions made a huge difference in how difficult the situation would ultimately become. Against the advice of federal authorities, the mayor of the City of New Orleans did not issue an evacuation order until it was too late to get everybody out, and poorly coordinated the manpower and transportation resources needed to evacuate those who did not have transportation. These decisions had significant repercussions on death rates and on the well-being of those who were forced to stay behind. Once the effects of the storm struck the city, the authorities were no longer capable of mounting an effective response due to the vast infrastructure damage and the evacuation of a large number of their government employees. In the immediate aftermath of the hurricane, New Orleans became dependent upon state- and federal-level resources, as did many other coastal and near-coastal communities in Louisiana, Mississippi, Texas, and Alabama.

Katrina hit four states and some aspects of the federal response to all four were confused, at least initially. However, in Mississippi, Texas, and Alabama, the response coordination became increasingly better, whereas it did not in Louisiana. This demonstrates the importance of that middle level of response, the state, bringing about a well-coordinated response. A quick civics lesson also sheds some light on why Louisiana's response was so markedly different than those of the other three impacted states. Since Louisiana was a French territory before it was purchased by the United States, its state constitution is based on the French provincial form of government. After the French revolution, the architects of the French constitution purposely created a weak central federal government in an attempt to ensure that no one would ever again garner enough power to name themselves emperor or king.

So, what does this have to do with how Louisiana responded to Hurricane Katrina? Well, the Louisiana state constitution vests much power at the parish level and little with the state. The extent of this local power is seen in the fact that the chief executive of each parish in Louisiana is called the parish president. All other states have English-based constitutions, where almost all power is vested at the state level and authorities are delegated to local government via state statutes and regulations. For states to be effective as midlevel responders in a catastrophe, state law and regulations must allow state authorities to centralize event response to fill gaps in local response, to coordinate the distribution of limited supplies and resources based on regional priorities, and to coordinate with federal responders.

All four of the states that were directly impacted by Hurricane Katrina operated under a standard all-hazards emergency operations plan, at both state and local levels. One of the concepts of the all-hazards standardized NIMS approach is that it minimizes confusion as new resources come in from the outside to provide assistance. These additional resources all have the same authority structure and use the same vocabulary. However, they do not have the same response plans for the event type at hand, and this created significant problems in the response to Hurricane Katrina. The disarticulation between plans was not only between state and federal plans, but also between the various local jurisdictions within a given state. This was particularly problematic in Louisiana, and contributed to the mass response confusion that reigned in that state for some time. Another area of disarticulation was between federal agencies and, at times, between federal-level political operatives and emergency managers. The breakdown of the incident management system that resulted from political-level interventions led to such confusion that it sometimes took days, or even weeks to reconstitute organized command. For an overview, from the federal perspective, see the official evaluation released in 2006 by the White House and Congress (http://www.whitehouse.gov/reports/katrina-lessons-learned/). Note that the Government Accountability Office (GAO) has issued numerous reports on various aspects of the response to Hurricane Katrina. You can find this on the Internet at: http://www.gao.gov/docsearch/locate?searched=1&o=0&order_by=rel&search_type=publications&keyword=Hurricane+Katrina&Submit=Search.

Obviously, lessons have been learned from the response and recovery from Hurricanes Katrina and Rita, and are still being learned. Texas learned from watching the late evacuation of New Orleans and, when faced with Hurricane Rita a month later, initiated and operated a much more effective evacuation of vulnerable populations in eastern Texas.

FEMA learned that some kinds of hypercomplex events, the focus of this book, require multijurisdictional, multiagency, and multigovernment joint planning and exercising if a region is going to be able to have a coordinated response to predictable events. One of the lessons observed by researchers, such as Lagadec (2007), is that the vastness of the needs caused by the event, and the confusion caused by attempts at response, led to a level of complexity that was too much for the current U.S. emergency management response system to handle effectively.

Another observed characteristic of a catastrophe, such as Hurricane Katrina, is that they have long-term effects, one of which is that a catastrophe that results in widespread damage to homes, infrastructure, and the local and regional economies, will provoke significant out-migration of affected populations. Whether these populations return in significant numbers during the recovery and reconstruction phase is a direct result of how effective the response to the event is by government and non-governmental organizations (NGOs). Section III of this book addresses mass care and mass migration issues associated with catastrophes.

2.5.7 Haitian Earthquake

BOX 2.8 SITUATION: HAITIAN EARTHQUAKE

- The U.S. Geological Survey called it the strongest earthquake since 1770 in what is now Haiti.
- The quake struck on January 12, 2010, at 4:53 p.m.
- The 7.0 magnitude quake's epicenter was just 10 miles west of Port-au-Prince and its 2 million inhabitants.
- Three million people were in need of emergency aid after the major earthquake.
- Through January 31, 2010, there were 33 aftershocks ranging in magnitude from 4.2 to 5.9.
- Some 9,000 peacekeepers have been in Haiti since 2004, including 1,266 Brazilians. As of February 5, 2010, the United States had dispatched 13,000 service members to assist in the response.
- Haiti has no real construction standards, leaving their preferred form of hurricane-resistant construction (cast-in-place concrete and concrete block) highly vulnerable to collapse in an earthquake.

At the time this book was written in early 2010, the Haitian earthquake response had been underway for about a month. At that time, few, if any, assessments of the international response had been launched or completed. Nonetheless, it is certainly worth briefly exploring what has been learned about this catastrophic event in the 30 days since it occurred. At this time, the scale of the event, in terms of deaths and injuries, appears to be similar to the 2004 Indian Ocean tsunami in which approximately 240,000 people died. On January 28, 2010, the Haitian government gave a confirmed death toll of 170,000, with many more thousands dead in the rubble and outside the capital. This government estimate also did not include unreported bodies buried by relatives. The vast majority of casualties were Haitian civilians, but among the dead were aid workers, foreign embassy staff, foreign tourists, and a number of public figures including the archbishop of Port-au-Prince, Monsignor Joseph Serge Miot, officials in the Haitian government, and opposition leader Michel "Micha" Gaillard. Also killed were a number of well-known Haitian musicians and sports figures, including 30 members of the Fédération Haïtienne de Football, their national soccer team. At least 85 United Nations personnel, working in support of their peacekeeping and national building efforts, were killed, Among them were the mission chief, Hédi Annabi, and his deputy, Luiz Carlos da Costa (*Sydney Morning Herald*, 2010). As one can see, the losses from the earthquake tore across the whole fabric of Haitian society. In subsequent chapters of this book, we will explore how these types of catastrophic events impact the ability of a community to redefine itself and reestablish an identity that has been lost through the death of its political, cultural, and religious leadership and institutions.

With respect to the logistics aspects of the earthquake response, one significant difference between the two events (the Haitian earthquake on Jan. 12, 2010 and the Asian tsunami of Dec. 26, 2004) is the concentration of both damage and population observed in the Port-Au-Prince, Haiti, impact area. The concentration of population created both challenges and benefits. Challenges include having to deal with throngs or victims compressed into a small area, and little undamaged infrastructure available to support caring for large concentrations of both the injured and uninjured. Current estimates include 300,000 victims with injuries and 3 million significantly impacted persons that are in need of life-supporting assistance. Benefits include a relatively short logistics chain to reach the impacted population once supplies are in theater (Figure 2.6), the damage occurred within one country, making it somewhat easier to manage the political aspects of the response, and it is easier to see the full operational picture since the impacted area is composed of a relatively compact geographic area.

FIGURE 2.6 U.S. Navy Sea Hawk helicopters from the aircraft carrier USS Carl Vinson transport water and supplies from the airport to areas around Port-au-Prince, Haiti, January 18, 2010. (U.S. Navy.)

2.6 POTENTIAL FUTURE CATASTROPHES

2.6.1 Sea Level Rise

BOX 2.9 SITUATION: SEA LEVEL RISE

- With climate change taking place, there is significant melt off of Arctic, Antarctic, and terrestrial glaciers, with resulting waters going to the oceans.
- Estimates range between 1 and 3 m sea level rise in the next century, but current melt offs are progressing much more quickly than any model had predicted.

We now move to discussing potential future catastrophes, with focus on only a few of the numerous potential events:

- Familiarize the reader with some of the event types whose causal factors are already in progress and visible.
- Get the reader used to thinking of catastrophes that have a continuum, including, for many, a lead time that allows for mitigation and preparedness.

- Expose the reader to the reality that hypercomplex events also have hypercomplex etiologies, and that understanding those etiologies provides opportunities for intervention at various points, both pre- and postevent onset.

Sea level rise is an excellent example of a slow onset event, which we can see now underway in the daily news. Sea level rise is both a direct and indirect threat. This discussion addresses the two direct threats: (1) the ocean will flood costal areas that are currently, in some cases, densely populated, and (2) it will contaminate the fresh groundwater of many areas that are close to the new coastline, but not yet flooded with saltwater. In the first case, populations will have to migrate to higher ground; in the second, they will lose significant agricultural productivity and drinking water availability in a world that is already moving toward food and drinking water inadequacy. Island populations will be particularly hard hit, as many of them are low-lying and have nowhere on island to which they can relocate. If authorities can manage peaceful out-migration to another country, the loss of the island might be a significant inconvenience, but if peaceful migration doesn't occur, the situation could become catastrophic. See the cases of two island countries (Tuvalu and the Maldives) that, at the time this book was written, were considering the need to evacuate their entire populations to another country, due to rising sea levels. More than 300 million people are estimated to live within the 1-m (3+-ft) sea level rise inundation zone. Note that many European and international planning agencies are now setting their planning targets based on the assumption of a 5-m (15+-ft) sea level rise (Dasgupta et al., 2007; Olsthoorn et al., 2005).

An indirect result of sea level rise is that storm surges will have a deeper reach into populated coastal areas. This comes at a time when climatologists are predicting increasingly powerful storms. This raises the fear that Katrina-like events could become more commonplace. Either way, mass population relocation will be an inevitable result if the sea level continues rising, and it won't just be people moving around within their own countries. As we've already mentioned, mass migration is much more likely to result from catastrophes than disasters, and such migrations, in and of themselves, have the possibility to become a secondary catastrophe. There are very few examples of mass migrations in the past 500 years of human history that have been accomplished peacefully. We will explore this important characteristic of catastrophes further in Section III.

Much valuable agricultural land around the world lies in river deltas, often within a couple of meters above mean high tide level. This is particularly true in places like Bangladesh, India, Thailand, Cambodia, Myanmar, China, and even the Chesapeake Bay area of the United

States. In many of these areas, the agricultural production from these delta plains is important to the survival of populations that are already at risk of starvation. Sea level rise will begin to ruin agriculture by way of inundation for those on or near the coastline, and also can affect agriculture farther inland by way of contaminating aquifers with saltwater that are used for crop irrigation. Such is the predicted case with low-lying Florida, whose produce is sent around the world (Reese, 2000).

2.6.2 Drought and Desertification

BOX 2.10 SITUATION: DROUGHT AND DESERTIFICATION

- A combination of climate change and poor land use practices is resulting in loss of overall fresh water available in some significant parts of the world.
- This is causing loss of land capable of supporting either crops or natural vegetation.
- Without vegetation, land converts to desert.

This is another natural event that can be seen coming; the process is already underway and is well documented. Photos from sites of new desertification can really help you see this as a reality (Figure 2.7). For starters, go to http://www.abc.net.au/science/photos/desertification/ *or* http://managingwholes.com/photos/index.htm.

This is a problem in many parts of the world and, if Darfur is any example of the human response to drought and desertification, conflicts over increasingly scarce, arable land may prove as dangerous as starvation. The rapid loss of underground aquifers adds to the consequences of drought, in many places removing future options to irrigate land that is dry on the surface. If the current drought in the American Southwest continues unabated, it could lead to large-scale relocation.

Some of the numbers that help define the global extent of drought and desertification with los of arable land to desert per year (Brown, 2008):

China: 3,600 km^2 (roughly the size of the state of Rhode Island)
Nigeria: 3,510 km^2

From a global perspective, climate change-generated drought in the American Midwest could prove far more catastrophic than the loss of Lake Mead in the Southwest because hundreds of millions of people around the world directly or indirectly depend on grains from

FIGURE 2.7 Landsat photo of dust storm across Pakistan. In the last 25 years satellite imagery has helped to increase our understanding of desertification, indicating how areas change over progressive seasons and the potential impact to people and animals. However, other types of remote sensing, ground-based observances, and a global network of databases is needed before desertification is fully understood. (NASA photo.)

the "American Breadbasket," even in normal times. If climate change continues to decrease crop yield in many parts of the world, each source of grains will become that much more important.

By this point it should be becoming apparent that the hazards and threats we are talking about here may originate years before reaching catastrophic proportions, and may have an etiology that originates on other continents. A drought in North America's Breadbasket can result in many thousands, if not millions of deaths in Asia and Africa. Year-after-year loss of snow pack in Colorado, the source of much of the irrigation and drinking water west of the Rockies, can result in millions of people in Southern California migrating to places with a more secure water supply. Economists have traditionally assured us that if there is a demand, somebody will figure out how to supply the goods demanded. That is clearly not the case where the basic resources we call on are in short supply. Remember that during the Irish potato famine, the British government took a laissez faire approach with tragic results. These are not just ecological issues; they are the progenitors of potential catastrophes. What skills will emergency managers need in order to contribute to efforts to minimize the effects of slow-moving, but massive, hazards like drought and desertification?

2.6.3 Global Pandemic

BOX 2.11 SITUATION: GLOBAL PANDEMIC

- A pathogenic virus or other microbe mutates (either naturally or by purposeful weaponization) and spreads out into a human population that has no natural immunity to it.
- There is a significant lag time between the spread of the disease and the ability of medical and pharmaceutical researchers to develop and produce enough vaccines or medicines to effectively counteract the microbe.

The scenario of a pandemic presents a real challenge to practicing emergency managers. This is partly because emergency managers tend to have minimal understanding of the health science behind disease control. The scenario also runs counter to the more frequent disaster events, in which some outside force damages physical structures and hurts humans in the process. In the pandemic situation, the physical infrastructure is not destroyed, but humans are the locus of the damage.

One of the challenges in a pandemic is that the functioning of the human-built organizational infrastructure will be severely affected, including perhaps some basics like the transport and distribution of food and other basic commodities. The medical system we depend on to take care of the sick will be rapidly overwhelmed, making it necessary for authorities to make really difficult choices. It is not inconceivable that even basic public safety and security will become challenged.

2.6.4 New Madrid Mega-Earthquake

BOX 2.12 SITUATION: NEW MADRID MEGA-EARTHQUAKE

- The New Madrid fault line along the Mississippi River repeats its 1811–1812 production of one or two Richter-level 8 earthquakes.

The New Madrid Seismic Zone major earthquake scenario is one of FEMA's major planning points for catastrophic preparedness in the United States and has been vetted from many angles as being quite realistic. What converts this event into a catastrophe is the combination of the size of the event (magnitude), the lack of appropriate building codes and building designs, and the important roles the area plays in the national infrastructure and economy. As is the case with all disasters, this event becomes a catastrophe because of the vulnerability humans built into the built environment, private and public, as well as the design of service systems in a way that assumes their immunity from seismic hazards.

The core message in this discussion is that the damage and disruption would be so widespread that:

1. The entire country would be affected.
2. Local mutual aid compacts would be of little use. Response would require a different kind of thinking and unconventional approaches to logistics.

Many places in the world depend on grains transported down the Mississippi River from the American Breadbasket. The sheer quantity of the damage would render rapid reconstruction impossible. The Mississippi River would probably remain unnavigable for an extended period of time and the east–west road and rail network would be cut as river crossings would be severely damaged. In addition, probable significant out-migration from the impacted areas would occur.

From this discussion, we see some of the core characteristics of a catastrophe: health impacts that are well beyond regional capabilities, economic impacts that have both national and international significant consequences, and damage that is so severe that people migrate out of the region in order to survive. Now is the time to start looking at this scenario from the perspective of *all* the resultant needs that would have to be addressed, ranging from basic survival resource provision, to healthcare, social services, monetary and economic systems, law enforcement, infrastructure restoration (including transportation and energy supplies), and, among everything else, prioritization (Figure 2.8).

2.7 FACTORS COMMON IN CATASTROPHES

Catastrophic events will almost certainly involve many jurisdictions simultaneously. Jurisdictions are unable to respond effectively alone, but the breadth of the event makes outside help difficult or impossible, at least in the early days after a no-notice event. Response demands will outstrip the capabilities of traditional government leadership and resources. Remember, catastrophes can occur with no warning (earthquakes), some warning (hurricanes, typhoons, cyclones, major flooding), and considerable warning (droughts).

With this summary of factors common in catastrophes, let's go back up to the conceptual level. This summary should help the reader

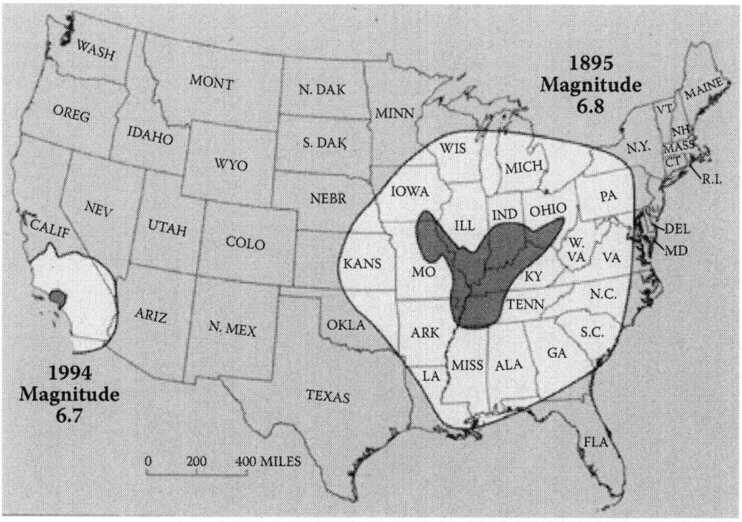

FIGURE 2.8 New Madrid earthquake map. The two areas shown were affected by earthquakes of similar magnitude: the 1895 Charleston, Missouri, earthquake in the New Madrid seismic zone and the 1994 Northridge, California, earthquake. (USGS graphic.)

internalize your own working definition of "catastrophe," and begin to realize what a gargantuan task it is to prepare for and respond to such events.

Perhaps the most important two points of this discussion are the ones that break the stereotypes of sufficient government leadership in any kind of event, and that all high-impact events are the rapid onset type.

As the reader summarizes this discussion, they should clearly see the need for emergency management, as a profession, to start thinking and working well outside of the parameters that have become "normal" in the past 25 years or so. It should now be clear to the reader, if we are going to be useful in protecting our populations in the face of upcoming catastrophes. It is important to note that:

- The complicated etiologies of most causes of catastrophes require strong and sustained interaction between science and emergency management in order to provide any chance of successful catastrophic mitigation or response planning.
- Poor government response, such as the Irish potato famine or, more recently, Hurricane Katrina, can significantly enhance the peril.

The first point above cannot be overemphasized. If emergency managers are going to serve a role of bringing order to hypercomplex events, they will need to be conversant with at least a basic level of the sciences that inform us of the etiologies of the phenomena that lead to catastrophes. The second point seems self-evident, but is needed anyway, making sure that the reader recognize that we cannot expect good results to come out of poor response management.

2.8 DISCUSSION QUESTIONS

1. What difference does it make whether you consider an event a disaster or a catastrophe?
2. Assume Lake Mead goes dry in 2021 due to climate change and prolonged drought in the Colorado Basin area. About 22 million people in three states depend on water from that one source for survival. How should emergency managers and other government officials in Las Vegas prepare to respond?
3. What could European and North American governments and other organizations have done to decrease the suffering and deaths during the European Little Ice Age?
4. How did government decisions during the Irish potato famine worsen the situation? What good and poor decisions did the affected population make?
5. Massive population migrations will likely result from sea level rise and drought/desertification. What role might emergency management play in minimizing pain and tragedy when large populations relocate? What barriers will need to be addressed?
6. How can the work of the emergency management and public health communities complement each other during a catastrophe?

REFERENCES

BBC News. 2005. Useless teddy bear—Kadirgamar. March 25. Web site: http://www.bbc.co.uk/sinhala/news/story/2005/03/050317_laxman_london.shtml (accessed January 10, 2010).

Benedictow, O.J. 2004. *The black death 1346–1353: The complete history*. Rochester, NY: Boydell & Brewer.

Brown, L. 2008. *Plan B 3.0 mobilizing to save civilization.* New York: W.W. Norton & Company.

Columbia Encyclopedia, The 2009. New York: Columbia University Press.

Dasgupta, S., B. Laplante, C. Maisner, D. Wheeler, D.F. Yan. 2007. The impact of sea level rise on developing countries: A comparative analysis. World Bank Policy Research Working Paper No. 4136. Online at http://papers.ssm.com/so13/papers.cfm?abstract_id=962790 (accessed April 24, 2010).

Fagan, B. 2000. *The little ice age: How climate made history*. New York: Basic Books.

Fox News. 2005. Indonesia reasserts control over aid. January 13. Online at: http://www.foxnews.com/story/0,2933,144109,00.html (accessed January 10, 2010).

Heyman, D.L. ed. 2008. *Control of communicable diseases manual.* Washington, D.C.: APHA Press.

Intergovernmental Panel on Climate Change (3rd assessment). 2001. Was there a "Little Ice Age" and a "Medieval Warm Period"? Working Group 1: The Scientific Basis. Chap. 2: Observed climate variability and change. Online at: http://www.grida.no/publications/other/ipcc_tar/?src=/climate/ipcc_tar/wg1/070.htm (accessed January 10, 2010).

Johnson, N., and J. Mueller. 2002. Updating the accounts: Global mortality of the 1918–1920 "Spanish" influenza pandemic. *Bulletin of the History of Medicine* 76(1):105–115.

Kolata, G. 2001. *Flu: The story of the great influenza pandemic of 1918 and the search for the virus that caused it.* New York: Simon and Schuster.

Lagadec, E. 2007. *Unconventional crises, unconventional responses: Reforming leadership in the age of catastrophic crises and hyper-complexity.* Baltimore, MD: Johns Hopkins University Press.

Margesson, R. 2008. CRS report for Congress. Indian Ocean earthquake and tsunami: Humanitarian assistance and relief operations updated February 10, 2005. Online at: http://fas.org/sgp/crs/row/RL32715.pdf (accessed April 11, 2008).

Markel, H., H.B. Lipman, J.A. Navarro, A. Sloan, J.R. Michelsen, A.M. Stern, and M.S. Cetron. 2007. Nonpharmaceutical interventions implemented by US cities during the 1918–1919 influenza pandemic. *JAMA* 298(19):2260.

NASA. 2001. The Sun's chilly impact on Earth. Scientific Visualization Studio, Goodard Spaceflight Center, Greenbelt, Maryland. online at http://svs.gsfc.nasa.gov/stories/iceage_20011207/ (accessed January 10, 2010).

Olsthoorn, X., P. van der Werff, L. Bouwer, and D. Huitema. 2005. Neo Atlantis: Dutch response to five meter sea level rise. Working Paper FNU75. Institute for Environmental Studies, Vrije Universiteit Amsterdam. Available at: http://www.uni-hamburg.de/Wiss/FB/15/Sustainability/waishollandwp.pdf (accessed January 10, 2010).

Orent, W. 2004. *Plague: The mysterious past and terrifying future of the world's most dangerous disease*. New York: Free Press.

Reese, R.S. 2000. *Hydrogeology and the distribution of salinity in the Florida aquifer system, Southwestern Florida*. Washington, D.C.: U.S. Department of the Interior, U.S. Geological Survey.

Reuters News Service. 2004. Jan Egeland news conference, December 27, 2004.

Schwartz, P. and D. Randall. 2003. *An abrupt climate change scenario and its implications for United States national security*, Washington, D.C.: U.S. Department of Defense.

Science Daily. 2008. Lake Mead, key water source for Southwestern US, could be dry by 2021. Available at: http://www.sciencedaily.com/releases/2008/02/080212141424.htm (accessed January 10, 2010).

Scripps Howard News Service. 2008. Lake Mead could be dry by 2021. February 13. Available at: http://scrippsnews.ucsd.edu/Releases/?releaseID=876 (accessed January 10, 2010).

Sydney Morning Herald, The. 2010. Eighty-three UN staffers killed in Haiti (2010). Agence France-Presse, January 28. Online at: http://news.smh.com.au/breaking-news-world/eightythree-un-staffers-killed-in-haiti-20100128-n00t.html (accessed January 28, 2010).

Tsunami Evaluation Coalition. (2006). Coordination of International Humanitarian Assistance in Tsunmai-Affected Countries. Retreived January 10, 2010 from http://www.alnap.org/initiatives/tec.aspx

United Nations Secretary-General's Statement, The. 2005. January 3, 2005. Online at: http://www.un.org/apps/sg/sgstats.asp?nid=1247 (accessed January 10, 2010).

Washington Times. 2004. U.N. official (undersecretary-general for humanitarian affairs Jan Egeland) slams U.S. as "stingy" over aid. December 27.

I Conclusion

Because it is important to understand the unique characteristics of catastrophes, in Section I we reviewed the history of catastrophic events both in and outside the United States, how catastrophes differ from disasters and emergencies, and how they are all part of the emergency management continuum. We also explored the varying definitions of catastrophes and their political and societal implications.

We explored several definitions of *catastrophe*, and looking at these varying definitions gave us insight into how differing communities, such as academicians versus practitioners, interpret the definitions. Information was presented concerning famed disaster sociologist E. L. Quarantelli's (2006) list of six criteria that help distinguish disasters from catastrophes. His definition included indicators that are recognized at the community level as well as addressing national actors. Quarantelli's definition has become well used in the limited but growing sociological literature on catastrophes. The *Continuum of Magnitude* offered additional insight into how to understand the difference between a disaster and catastrophe. The Continuum of Magnitude ranges from an emergency that can be managed using locally available resources to a so-called extinction level event that can't be managed at all.

An introduction to catastrophic planning was presented, including a brief discussion that introduced the reader to the challenges of attempting to apply disaster planning principles to catastrophic planning. We learned that multihazard-based planning, used for disaster planning, often doesn't apply to planning for a catastrophe. Section IV of this book will explore, in greater depth, planning issues associated with responding, recovering, and reconstructing after a catastrophe.

To place the reader in the right frame of mind, brief examples of historical catastrophes were presented that illustrated that catastrophes have occurred throughout history, have been caused by both manmade and natural causes, and that human decisions have often played a significant role in the severity of the impact of the event. With this understanding, we then briefly assessed the potential effects a catastrophe might have on the U.S. emergency management system.

Finally, we explored potential future catastrophes and the factors common in such events. Section I of this book culminated in a look at past and future catastrophes—their etiologies and challenges.

SECTION II

Ethical, Political, and Legal Issues

- Overview
  Section II of this book is designed to introduce the inherent ethical, political, and legal issues of catastrophic events. With respect to ethics, we will address the main ethical and value dilemmas and quandaries that will likely be faced before, during, and after a catastrophe. This part of the book also explores the legal framework associated with government response to catastrophes that includes the use of the military for domestic response, suspension of civil rights, and federal control of industrial output. Political factors, as well as organizational dynamics, are included to provide a basis for understanding the complex environment in which preparing for extreme events may take place. The body of literature reflected in political science and public administration provides us with insights into conflicts that arise during highly stressful events and the nature of the problems that evolve from our attempts to deal with catastrophic level events.
- Learning Objectives — By the end of Section II, the reader should be able to:
  - Analyze major ethical issues that arise in the course of catastrophic events; to identify and discuss the major ethical quandaries and dilemmas faced in preparing for and responding to a catastrophic event.
  - Apply basic tools and techniques of ethical analysis to one or more catastrophic events, and to glean insights about

morally appropriate and ethical response actions and decisions.
- Appraise how the National Response Framework, the National Preparedness Guidelines, and the National Incident Management System shape the readiness and response of the United States for extreme events.
- Compare the principles of federalism with the exceptions to federalism (e.g., major exceptions: Posse comitatus, use of military forces, police roles of military forces, etc.
- Critique current federal government plans for catastrophe readiness and response and potential federal system breakdowns in potential future catastrophic events.
- Appraise the political structure of the U.S. emergency management system, and the use of the National Guard, Coast Guard, and other military forces for domestic support in catastrophe response.
- Analyze government legal powers during catastrophes, including state legal protection laws for volunteers.
- Analyze ways in which political and legal change can result from catastrophes.
- Critique the challenges of interjurisdictional partnership issues.

- Outline of Topics
  - Chapter 3: Ethics
    - An Introduction to Catastrophic Ethics
    - Defining Ethics
    - Ethical Duties Related to Professional Roles
    - The Moral Community: How Is it Defined in Terms of Catastrophic Response and Readiness?
    - Competing Ethical Theories and Frameworks
    - Utilitarian Catastrophic Response
    - Deontological Perspectives; Duties and Principles to Govern Catastrophic Planning and Response
    - Environmental Ethics
    - Virtue Ethics and Catastrophic Response
    - NGO Ethical Dilemmas
  - Chapter 4: Political and Legal Issues
    - FEMA Comprehensive Preparedness Guide
    - The National Incident Management System (NIMS)
    - The National Response Framework (NRF)
    - Principles of Federalism and Exceptions to Federalism
    - Intergovernmental Collaboration: A Cornerstone of Effective Catastrophic Planning and Response

- Federal Government Plans for Catastrophe Readiness and Response Ensuring Enduring Federal and State Constitutional Governments
- Local Government Emergency Response Plans
- Political Structure of the U.S. Emergency Management System
- Use of the National Guard, Coast Guard, or Other Military Forces in Catastrophe Response
- Potential Federal System Breakdowns in Hypothetical Future Catastrophic Events
- State Government Legal Powers during Catastrophes
- State Legal Protection Laws for Volunteers
- Political Implications of Catastrophes at Various Governmental and Political Levels
- Political and Legal Change That Can Result from Catastrophes
- Challenges of Interjurisdictional Partnerships

- Section II Conclusion

CHAPTER 3

Ethics

3.1 LEARNING OBJECTIVES

At the end of this chapter, the reader should be able to:

- Analyze major ethical issues that arise in the course of catastrophic events; to identify and discuss the major ethical quandaries and dilemmas faced in preparing for and responding to a major natural disaster or catastrophic event.
- Apply basic tools and techniques of ethical analysis to one or more catastrophic events, and to glean insights about morally appropriate and ethical responses, actions, and decisions.
- Analyze government legal powers during catastrophes, including state legal protection laws for volunteers.
- Analyze ways in which political and legal change can result from catastrophes.

3.2 KEY TERMS AND PHRASES

- Anthropocentrism
- Biocentrism
- Deep Ecology
- Deontological theory
- Ecocentrism
- Ethical holism
- Environmental Ethics
- Ethics
- Moral Community
  - Biological Dimension
  - Temporal Dimension
  - Geographical Dimension
- Nongovernment Organization Ethical Dilemmas
- Paternalism

- Professional Duties
- Role of Rights
- Teleological theory
- Theory of Justice
- Utilitarian Catastrophic Response
- Utilitarian Environmental Ethics
- Virtue Ethics

3.3 AN INTRODUCTION TO CATASTROPHIC ETHICS

In this chapter, we will explore the answers to the following questions:

- What are ethics?
- From where do they derive?
- What's their relevance to catastrophic events?

Many believe that at the heart of what makes human beings special or unique is the ability to engage in ethical and moral reasoning, and to act and live according to ethical principles. The goal of a good and meaningful life can be argued. Some would say it is living a principled life.

That we are profoundly moral creatures and that we aspire to be principle bound does not mean that there is a clear or easy path to living an ethical life or to making an ethically correct policy or political decision. There will be (and are) fundamental disagreements about the content of values and ethics, and, to be sure, many obstacles (e.g., psychological, social) to making ethically defensible and ethically neutral decisions exist (e.g., as opposed to decisions based on personal self-interest).

Ethics are important to catastrophic planning and response because an emergency manager may be called upon to make major decisions that involve the health, safety, and welfare of those impacted by the event that an emergency manager wouldn't face during less severe events, such as an emergency or disaster. Remember, in a postcatastrophic environment, an emergency manager may find his/herself in a strikingly different world where many of the basic assumptions of a functional modern society no longer exist. So, how does one make such critical decisions in such an unfamiliar setting? Emergency managers fall back on their moral values and ethics in conjunction with legal considerations to serve as a compass to steer them into making the sound decisions or at least making the best decision from a series of bad choices, which is often the choice an emergency manager has available after a catastrophe.

As a species, we continue to debate where and how we derive our moral sensibilities. For many, there are important religious sources of morality and ethics (e.g., the Bible and other religious scripture). Some biologists, E. O. Wilson (1975) for instance, argue that moral standards and codes are borne of evolutionary need; ethics codes and standards provide important guidance for survival and cooperation.

Many moral philosophers commonly appeal to rationality in one form or another in order to defend the notion that it is moral principles that govern behavior. We have a need for and adopt moral principles because it is rational to do so (i.e., we would embrace and consent to such a moral framework). Gert (2004) makes this argument in his book *Common Morality* that examination of the content of common morality makes it clear that it is a system; that it would be rational for all persons to want everyone to be taught and trained to follow it because of the protection that it provides to themselves and for those for whom they are concerned. It is rational for all persons to want everyone to obey rules, such as "do not kill," "do not deceive," and "keep an emergency manager's promises," with regard to themselves and those for whom they are concerned.

John Rawls (1971), in his classic treatise *A Theory of Justice,* argues for a particular set of principles of justice based on a thought experiment: Under conditions of neutrality, under a "veil of ignorance" about one's own personal economic and social circumstances, what would rational, risk-averse individuals accept and accede to? Rawls' two principles of justice, then, are derived, grounded, and defended based on the rational deliberation of humans. Rawls argues that rational individuals under these fair conditions of neutrality will adopt this particular set of ethical principles (i.e., divorced of specific personal information that might bias the decision, say, that one is wealthy or holds an especially high social status). The principles of justice are not defended based on intuition or by reference to religion or some other a priori moral standard, but on human logic and rationality.

Ethics and moral philosophy, though a long-standing academic discipline and area of study, should not be viewed as providing absolute answers or perfect clarity of decision making in a catastrophic setting. Rather, it is as much about a process of searching, probing, asking questions, and an earnest effort to consider the range and significance of the moral issues involved. Ethics is as much an endeavor, an ongoing journey, in which one identifies the moral concepts and factors relevant to each specific case or dilemma, and works in good faith to reach ethical judgments that are at least thoughtful and informed.

3.4 DEFINING ETHICS

What are ethics and what sorts of issues do they entail? An initial definition provided by the Josephson Institute for Ethics (Los Angeles) is:

> Ethics refers to standards of conduct, standards that indicate how one should behave based on moral duties and virtues, which themselves are derived from principles of right and wrong.
>
> **Josephson Institute, nd**
> **http://josephsoninstitute.org**

Ethics involves questions of right and wrong, good and bad. The questions dealt with by ethics are often said to be questions primarily about "ought" as opposed to questions of "is" or factual questions. What an emergency manager feels is the morally correct decision will be related to how an emergency manager answers these questions. For instance, whether government has a duty to take quick steps to rescue a community following a catastrophe will depend on the factual assessment of whether, and to what extent, real and immediate danger is present. This discussion of ethics will involve some of the following:

- Duties and responsibilities of individuals, groups, professions
- Rightness or wrongness of specific actions or policies
- Competing values and value systems
- Fairness and equity in distribution of goods and services in a resource-constrained catastrophic environment
- Ethical principles to guide catastrophic response and recovery

During and after a catastrophic event there are many specific actions, decisions, policies, and programs that will need to be addressed and which will likely raise significant and serious ethical concerns. There are, moreover, many places where ethical guidance, ethical principles and concepts are needed to provide guidance and direction.

Previously, we discussed Josephson's classic definition of ethics. It is again important to emphasize that ethics is about questions of "ought" (as opposed to "is"), questions of right and wrong, good and bad, compelling values. More specifically, ethics are about:

- A conscious effort to unearth and consider the range and variety of ethical and moral issues and points of view.
- A process of asking questions, identifying assumptions, considering different points of view, and reaching judgments about correct and appropriate actions or courses of action before, during, and after catastrophic events.

It is worth noting that ignoring the ethical dimensions of catastrophic events is neither possible nor appropriate. The ethical and moral quandaries are ever-present and often at the core of major planning and response decisions. Ethics, then, are not optional, but an essential element in any response planning process.

At this point in the discussion, some key points worth summarizing with respect to a catastrophe are:

- Ethical dimensions are extensive and unavoidable.
- Failure to unearth and address the ethical quandaries does not make them go away.
- Ethics is more about a process than a definitive conclusion.

Section I of this book oriented the reader to the types of catastrophic circumstances that will likely be encountered. We will now look to identify the variety of possible ethical quandaries that will likely emerge from these circumstances. Here, the main point is to begin to identify some of the ways in which catastrophic events raise ethical issues and the nature of ethic dilemmas associated with response decisions.

At this point, readers will begin their orientation to the language of ethics as it deals with such issues as duties and responsibilities, questions about fairness and equity in the distributions of resources and aid, principle-based or guided judgments and actions, and the reconciling of sometimes conflicting values.

BOX 3.1 LANGUAGE OF ETHICS

- *Duties* of groups, professions
- *Rightness or wrongness* of specific actions or policies
- *Competing values and value systems*
- *Fairness and equity* in distribution of goods and services
- *Ethical principles to guide catastrophic response and recovery*
- *Responsibilities* of individuals

It is important to point out that the examination of the ethics of catastrophic events is a form of "applied ethics." The concern here is with understanding the ethical assumptions and providing tangible ethical and value guidance to the many serious and concrete response and recovery decisions that will need to be faced during and after a catastrophe.

All planning, policy, and management decisions, including those relating to emergency management and catastrophe response, involve ethical questions and quandaries. It is the overall goal of this discussion to convey to the reader the ubiquitous nature of ethical dilemmas and ethical assumptions, and to raise his/her overall awareness to them in catastrophe planning and response.

Catastrophic events raise the potential for many serious ethical dilemmas and quandaries that include:

- Distribution of benefits and burdens before, during, and following a catastrophic event.
- Questions of responsibilities and duties in catastrophic response, such as who is responsible for responding and what types of responses are required?
- Fairness of the unintended consequences.
- How much consultation/participation with the public and individuals affected, or potentially affected by response actions, is ethically required, especially when an emergency manager may have no means of communicating with the public following a catastrophe?
- Private versus public duties and responsibilities, i.e., government versus private and nonprofit sectors.
- How much responsibility should be attached to individuals and families to care for themselves following a catastrophe?
- Questions about the extent and nature of professional duties and responsibilities (e.g., do emergency management personnel have a special duty to attempt rescue even when doing so might put their own life at risk?).

This list of potential ethics issues is meant to be illustrative, but not exhaustive, of some of the ethical quandaries that emergency managers will confront in a catastrophe. It provides a few of the more common ethical dilemmas likely to emerge. More narrowly focused, the types of emergency management actions or policy issues that arise in a catastrophe that would raise serious ethical concerns include:

- Curfew, quarantine, and confinement: Can the rights and freedom and autonomy of individuals be superseded on behalf of the health and safety of the larger public? Restrictions on rebuilding in high hazard zones. Allocation decisions about the distribution and prioritization of postdisaster aid or medical treatment.
- Restrictions on civil rights and due process (e.g., in emergency conditions, can due process be suspended or modified or other constitutional rights suspended if necessary?). Limitations on personal and family movement (e.g., can families be prevented from returning to their homes and neighborhoods or from exercising their property rights when the conditions are unsafe?).

The above list provides a complementary but somewhat different take on this question, providing a few examples (again illustrative, not exhaustive) of actions or policies that might be commonly adopted and implemented before, during, or after a catastrophic event that raise serious ethical concerns or issues. Ethical issues will emerge, for instance, around the fairness of the distribution of resources (e.g., life-sustaining assistance) following a catastrophic event, and decisions about which neighborhoods, districts, groups, etc. receive public services (restoration of services, such as trash pick up and utilities), disaster assistance, or emergency services first.

Other issues entail questions of who is responsible for public safety in catastrophic event response (e.g., is it individuals and families primarily? Or, is it government? Is it the professionals of various sorts involved in one aspect or another of designing or preparing for, or responding to a catastrophe that are largely responsible? Or, is it all of these groups in some shape or form?).

There are many general beginning points about what ethics do and don't entail. The list below details some of the more important points, e.g., ethics are not optional, ethics are not like solving a mathematical equation, and ethics are different and distinct from law (taken up later as well). However, one's knowledge of legal ramifications may affect their ethical choices.

BOX 3.2 ETHICS: PRELIMINARY POINTS

- Catastrophic events raise many serious ethical quandaries.
- Ethical choice has a ubiquitous nature.
- Ethics are not optional, *de facto* ethical choices abound.
- Ethical decisions are not like solving an equation.
- Ethics are not the same as legality.
- Moral pluralism versus unitary moral theories.
- Values, attitudes, beliefs versus ethics.
- There are professional ethics, personal ethics, and policy ethics, among others.

There are many different sources of ethical insights, many different points of reference in making ethical judgments. These include intuition, theories, arguments, and thought experiments (e.g., consider John Rawls' (1971) description of the *Original Position*, used to argue for a particular set of principles of justice in *A Theory of Justice*), ethical codes, state or federal constitutions, and religious traditions, among others.

3.5 ETHICAL DUTIES RELATED TO PROFESSIONAL ROLES

There are many ethical issues as well as many questions about ethical duties and responsibilities that center on the professional roles and duties of individuals involved in managing and responding to catastrophic events. There are many different professions and professional roles involved and a few of these are highlighted. Professional ethics is a special category. Which fields, disciplines, or professions will have special ethical duties or responsibilities before, during, or following a catastrophic event? Some of these include:

- Architects, engineers, planners
- Emergency managers
- Doctors, nurses, other healthcare professionals
- Private industry and nongovernmental organizations (NGOs) responders.

There are codes of ethics and professional conduct that apply for each of these professions. Since emergency managers are not licensed professionals, like architects, engineers, and doctors, do emergency managers, as professionals, have special duties? The International Association of Emergency Managers (IAEM), an organization dedicated to promoting the goals of saving lives and protecting property by mitigating, preparing for, responding to, and recovering from disasters/emergencies, believes they do. IAEM sponsors the Certified Emergency Manager® (CEM) and Associate Emergency Manager (AEM) programs to maintain professionalism through the certification process. The "IAEM Code of Ethics and Professional Conduct" must be embraced and upheld by all individuals who are awarded the CEM/AEM designation (see Box 3.3). The promise to uphold the code signifies the assumption that the emergency manager will act prudently and responsibly beyond the requirements of law and codes. The Code of Ethics and Professional Conduct embodies the certification program philosophy and objectives. Each CEM/AEM promises to adhere to the code. For more information on the IAEM, see: http://www.iaem.com/about/IAEMCodeofEthics.htm.

BOX 3.3 IAEM CODE OF ETHICS AND PROFESSIONAL CONDUCT

Preamble

Maintenance of public trust and confidence is central to the effectiveness of the emergency management profession. The members of the International Association of Emergency Managers (IAEM) adhere to the highest standards of ethical and professional conduct. This Code of Ethics and Professional Conduct for IAEM members reflects the spirit and proper conduct dictated by the conscience of society and commitment to the well-being of all. The members abide by the association's core values of respect, commitment, and professionalism.

Values

Respect: Respect for supervising officials, colleagues, associates, and, most importantly, for the people we serve is the standard for IAEM members. We comply with all laws and regulations applicable to our purpose and position, and responsibly and impartially apply them to all concerned. We respect fiscal resources by evaluating organizational decisions to provide the best service or product at a minimal cost without sacrificing quality.

Commitment: IAEM members commit themselves to promoting decisions that engender trust and those we serve. We commit to continuous improvement by fairly administering the affairs of our positions, by fostering honest and trustworthy relationships, and by striving for impeccable accuracy and clarity in what we say or write. We commit to enhancing stewardship of resources and the caliber of service we deliver while striving to improve the quality of life in the community we serve.

Professionalism: IAEM is an organization that actively promotes professionalism to ensure public confidence in emergency management. Our reputations are built on the faithful discharge of our duties. Our professionalism is founded on education, safety, and protection of life and property. (From http://www.iaem.com/about/IAEMCodeofEthics.htm)

It is important to observe the distinction between law and ethics. An emergency manager will notice that the IAEM Code of Ethics and Professional Conduct calls upon CEMs and AEMs to "act prudently and responsibly beyond the requirements of *law* and codes." Laws are an expression of what is right, fair, and prudent. Law largely determines relationships and duties during a response. What is ethical and what is legal may not be the same. Laws relative to catastrophic response can and should change in response to ethical and moral discourse; after all, laws are rarely written in consideration of a future unknown catastrophe. They are, instead, often written to address past wrongs.

3.6 THE MORAL COMMUNITY: HOW IT IS DEFINED IN TERMS OF CATASTROPHIC RESPONSE AND READINESS

Considering and judging to whom we owe ethical obligations in a catastrophe is a significant and important question. To explore this further, we introduce the important philosophical concept of the *moral community*. The moral community refers to the people and things to which we might have ethical duties, and who it could be said share a community with us.

Once we define the moral community as people, places, and things to which we have some moral duty or responsibility to consider their interests and welfare, we must determine what additional duties or obligations or commitments we owe to them. For instance, if animals and other sentient (feeling) forms of life are part of our moral obligation, during and after a catastrophic event, do we have an affirmative obligation to rescue and care for them, or merely to treat them humanely when encountered in the aftermath of a catastrophe?

If future generations are part of our moral community, how far into the future are we obligated to take into account, and, specifically, what duties do we owe the future? There are many different ways to answer this question.

Some of the most important potential dimensions of the moral community include the following:

- *Biological dimension*: Are only humans part of our moral community, or do we have ethical duties to take into account the welfare and interests of other forms of life? If so, which ones? Are only sentient species (those that can feel pain) a part or are all life forms? Are only certain animals, say ones that we have actively cared for (i.e., households pets) owed respect and consideration? (*Note*: Rescuing and caring for pets during and after natural disasters is a major task and somewhat controversial when it takes away from time and resources that could be devoted to human rescue and care. What about animals who serve as a primary source of income?

- *Temporal dimension*: Are only residents living today a part of our moral community or do we also have ethical duties and responsibilities to future generations, and future residents? If we have a duty to consider impacts of our actions on the future, how far into this future does our moral community extend?
- *Geographical dimension*: What is the geographical range or extent of our moral community? Do we have duties to people and cultures far away, to respond to catastrophic events and impacts that take place in other nations, in other parts of the world? To what extent are those who are poor and vulnerable, but living in another country also to be considered part of our moral community and, thus, due respect and consideration?

3.7 COMPETING ETHICAL THEORIES AND FRAMEWORKS

Once we make some initial judgments about how we delimit or define the moral community, we must then consider the actual extent and content of the duties and responsibilities owed them. Much of this discussion will be devoted to considering and discussing the different ethical theories and positions that could serve to guide catastrophic response policy, actions, and decisions.

There are certainly different typologies, but one classic distinction in moral philosophy is between *teleological* theories of ethics and *deontological* ethics. The former bases the rightness or wrongness of an action or policy by assessing its consequences, and specifically by looking at the comparative balance of positive versus negative brought about. Utilitarianism is the dominant version of teleological ethics, and widely employed through tools, such as benefit cost analysis. Deontological ethics, on the other hand, justify rightness or wrongness by reference to some other ethical principle or standard than the comparative balance of consequences brought about.

Quoting University of Michigan philosophy professor William Frankena (1973), in his classic book *Ethics*:

- A **teleological** theory says that the basic or ultimate criterion of what is morally right, wrong, and obligatory, etc. is the non-moral value brought into being. The final appeal, directly or indirectly, must be to the comparative balance of good or evil.
- **Deontological** theory asserts that there are other considerations that make an action or rule right or obligatory besides the goodness or badness of its consequences, certain features of the act itself other than the *value* it brings into existence, for example, the fact that it keeps a promise, is just, or is commanded by God or the state.

Let's use a common ethics example to show the difference, say, promise-keeping. A utilitarian would judge the correctness of keeping his/her promises by assessing the comparative outcomes. If he/she keeps the promise, will more utility or benefit be brought about, than by ignoring the promise, and perhaps utilizing one's time in some other way? A deontological perspective would suggest that one has an a priori duty to keep a promise, that keeping one's promises is an ethical duty and should not be subject to utilitarian reasoning. A teleological perspective would judge the rightness or wrongness of keeping the promise by assessing the outcomes.

Beatley's (1994) two-dimensional graph of ethical positions is called the Topology of Ethics (Figure 3.1). The up–down axis is the utilitarian deontological, and a second, left–right axis, incorporates more recent thinking and debate about environmental ethics, whether there are duties to, or relative to, nature and natural environments (more on this below). The left–right axis (anthropocentric/nonanthropocentric) is an important one in the ever-expanding literature and thinking in environmental ethics. Many hold that decisions about whether and how to use the natural environment should be judged on essentially human-centered (anthropocentric) moral grounds, what will generate value for human beings, and under a utilitarian model, what will maximize the utility of those environmental "resources."

3.8 UTILITARIAN CATASTROPHIC RESPONSE

Box 3.4 and Box 3.5 provide some examples of the major ethical concepts and principles that fall within the teleological and deontological branches of ethics. While it is probably not possible to discuss in detail all of these ethical concepts and positions, it is important for the reader to appreciate that there are many more specific moral concepts here that fall (especially) within the deontological quadrants of the earlier diagram.

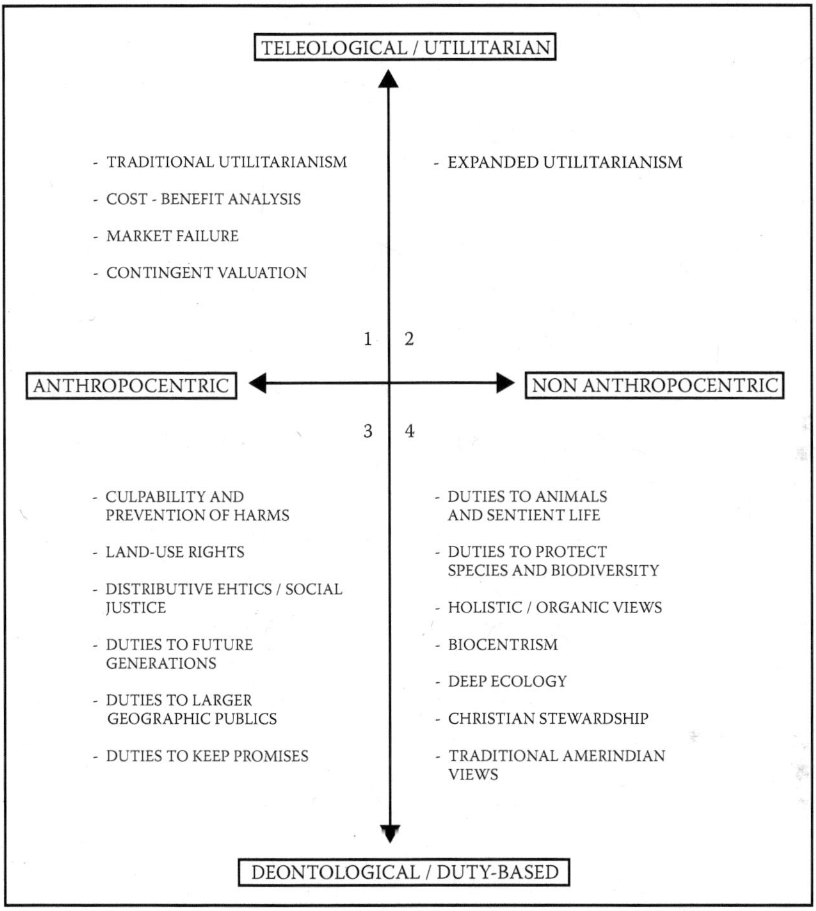

FIGURE 3.1 Beatley's Topology of Ethics. (From Beatley, T. 1994. *Ethical land use.* Baltimore, MD: Johns Hopkins University Press. With permission.)

BOX 3.4 TELEOLOGICAL/UTILITARIAN PERSPECTIVES

- Maximizing welfare/utility is the ultimate goal.
- Market values on land and environment are paramount.
- "Price" is the common metric; highest and best economic use.
- Value is determined through personal preferences and casting of dollar votes.
- Benefit cost analysis; contingent valuation.
- Present is given moral priority (e.g., discounting).

BOX 3.5 DEONTOLOGICAL PERSPECTIVES

- Individual rights and respect for personal autonomy
- Culpability and prevention of harm
- Social justice and equity
- Duties to future generations
- Duties to keep promises
- Duties to larger geographic publics/communities

Utilitarianism is the most prominent version of teleological ethical theory. It is grounded in a rich body of ethical theory and writings that emerged in the mid to late nineteenth century, including, most importantly, the work of John Stuart Mill (1863). In many ways, utilitarianism has emerged as the primary ethical underpinning for most contemporary planning and public policies. The idea of taking those actions that serve to maximize utility or public welfare, though, of course, there have been many disagreements about how to define or measure this, is the central insight. Much contemporary environmental and hazards/disaster policy reflects a strong utilitarian view.

Methodological techniques and tools, such as benefit cost analysis, are the most obvious manifestation of a utilitarian ethic, and have been incorporated in many emergency management programs and legislation. Benefit cost analyses, for instance, are mandated by law for the projects undertaken by the U.S. Army Corps of Engineers, such as hurricane protection and beach renourishment projects, and by FEMA for many of its emergency management grant programs offered to states, such as FEMA's Predisaster Mitigation Grant Program. There are many reasons to be critical of these analytic techniques (historically, these have been subject to much manipulation), but at the core is a strong utilitarian logic: For projects to move forward, they must show that the benefits exceed the costs, and those projects that maximize the ratio of benefits to costs should be preferred.

3.9 DEONTOLOGICAL PERSPECTIVES: DUTIES AND PRINCIPLES TO GOVERN CATASTROPHIC PLANNING AND RESPONSE

In general, nonutilitarian thinking and ethics govern most common morality. There is also a rich policy and planning history that establishes the moral basis for support on duty-based grounds. From air quality to endangered species protection to environmental justice, much contemporary environmental policy and planning finds its moral and ethical underpinnings in deontological theory.

The following discussion presents more detail about three specific (deontological) moral concepts that are especially important to contemplate in the context of catastrophic planning and response, those being individual autonomy, rights, and social justice.

3.9.1 Individual Autonomy

The classic philosophical definition of paternalism involves an action or set of actions taken and justified solely in reference to the benefit or welfare of the individual whose freedom is being constrained (Dworkin, 1972). Many actions and policies in natural disaster and hazard mitigation planning could be described in this way: For instance, restricting individuals from rebuilding, following a catastrophe, in very high risk environments (say a high risk hurricane or flood hazard zone, or volcanic eruption zone) where the negative impacts associated with such environments would fall essentially on the individual and not society (though, this is itself often a point of contention and debate). (See Beatley (1994) for a discussion of paternalism and land use planning.)

Many of these debates about acceptable risk and the specter of potentially paternalistic public action are influenced by the writing of John Stuart Mill (1859), principally in *On Liberty*. Mill defends the autonomy of individuals to make their own decisions—to craft their own life choices—with freedom from interference, unless an individual's actions have a public-regarding impact. Common public health-related restrictions on personal freedoms—e.g., the freedom to ride a motorcycle without a helmet—are justified on the grounds that there are indeed public costs and implications of these personal choices (high hospital and medical costs from brain injuries that fall to the public when accidents inevitably occur).

3.9.2 Rights

Let's now consider the moral concept of *rights*. What are the basic rights that individuals and groups in society are entitled to (by law or custom)? Do these rights serve as constraints on what society can do in response to a catastrophic event? A constitutionally protected right to private property, for instance, might prevent steps to clear debris, to clear and redevelop damaged areas, and to restrict the rebuilding of areas that are especially vulnerable to damage again from a future disaster event. Efforts at controlling crime and access to danger zones following a catastrophe might interfere with individual rights to free assembly.

BOX 3.6 THE ROLE OF RIGHTS

Rights are viewed as (mostly inviolable) moral constraints on public policy; rights place limits on benefit- or utility-maximizing actions.

Examples of some specific individual rights include:

- Political (e.g., right of free assembly, free press)
    - Legal or process (e.g., right to due process, habeas corpus)
    - Moral (e.g., right or entitlement to safe living conditions, adequate food, water, shelter, etc.)
- Rights are derived from, e.g., Constitution, UN Declaration on Human Rights, court opinions, long-standing custom or practice
- Rights can be moderated or curtailed (e.g., legitimate limits placed on free speech when that speech is hateful or dangerous; classic example of yelling fire in a crowded theater)

There are many different kinds of rights associated with catastrophes. One distinction often made in philosophy is between negative rights (right to be free **from** …) and positive rights (the right **to** ...). There are many potential sources from which to derive or argue for specific rights. The United Nation's Declaration of Human Rights is one such source that an emergency manager might find a useful resource (see United Nations (1948); the complete declaration can be found at: http://www.un.org/Overview/rights.html).

In the ethics literature, rights are often juxtaposed against utilitarianism. While taking steps and actions to maximize societal benefit or utility might be seen as desirable, a great challenge does not unnecessarily violate or undermine individual rights in doing so. Rights are usually viewed as trumping utilitarian policy.

3.9.3 Social Justice

With respect to the issue of social justice, there are many ways in which response to catastrophes might raise questions of social justice. For example, has there been profound inequality and inequity in the vulnerability of individuals and groups to future catastrophic events (e.g., there is evidence that minorities and the poor have greater vulnerability to hurricanes and coastal flooding), in the overall distribution of benefits and costs associated with catastrophes, and in the availability and access to resources and assistance following such an event? There are many different variants of social justice, and different definitions and interpretations. One commonly held ethical standard is that similarly situated individuals and groups should be similarly treated. As an emergency manager, this is an important goal in managing the consequences of a catastrophe.

Gert (2004) has argued that there is a common morality that is universal (or nearly so) that can be expressed in terms of 10 moral rules. While some or most of these rules could be defended or justified based on utilitarian reasoning, they are more typically thought to be duty-based or deontological (thus, falling more into Beatley's lower-left quadrant). Gert and others follow what has sometimes been referred to as "principlism," which is the idea that ethical and moral decisions can be effectively guided by adherence to these common principles or rules.

BOX 3.7 COMMON ETHICAL AND MORAL PRINCIPLES OR RULES

"Do not kill"
"Do not cause pain"
"Do not disable"
"Do not deceive"
"Do an emergency manager's duty"
"Obey the law"
"Do not cheat"
"Do not deprive of pleasure"
"Do not deprive of freedom"
"Keep an emergency manager's promises"

3.10 ENVIRONMENTAL ETHICS

During this discussion on environmental ethics, it is important to refer back to the *Topology of Ethics* diagram (see Figure 3.1). This discussion presents the most important recent positions and theories of environmental ethics from those most supportive of anthropocentrism to various nonanthropocentric positions. The nonanthropocentric views are interesting and almost always result in engaging, passionate discussion of all sides. *Utilitarian Environmental Ethics* (Baxter, 1974), *Biocentrism* (Taylor, 1986), *Ecocentrism* (Leopold, 1948), and *Deep Ecology* (Devall and Sessions, 2001) represent the more ambitious variants, and while each is quite different, they all assume a major ethical reorientation away from human-defined benefit and value, toward the intrinsic moral worth and inherent value of nature.

The following is a brief summary of some of the variants and ethical positions that might be useful in this discussion:

BOX 3.8 ENVIRONMENTAL ETHICS: PRELIMINARY DEFINITION #1

Environmental ethics is concerned with the issue of responsible personal conduct with respect to natural landscapes, resources, species, and nonhuman organisms (Partridge, nd). Arguably, environmental ethics represents one of the newest, most stimulating and exciting areas of ethical inquiry and discourse, and many catastrophic event responses directly relate to our natural environment and our ethical duties relative to it.

BOX 3.9 ENVIRONMENTAL ETHICS: PRELIMINARY DEFINITION #2

... concerned with the moral relations, which hold between humans and the natural world. The ethical principles governing those relations determine our duties, obligations, and responsibilities with regard to the Earth's natural environment and all the animals and plants that inhabit it (Taylor, 1986).

- *Utilitarian Environmental Ethics*: The value of a species or landscape determined by the stream of economic benefits generated from it. Moral value and worth are determined through a human value lens, often expressed through tools, such as benefit cost analysis and contingent valuation (Baxter, 1974).
- *Biocentrism*: The moral position and argument that all living things have a good of their own and, thus, inherent moral worth. Paul Taylor's (1986) book *Respect for Nature* remains the definitive work and lays out an elaborate set of principles and standards to guide actions that impact the natural environment.

- *Ecocentrism*: This position holds that it is not life per se that has moral worth, but complex ecosystems and larger landscapes, often framed in terms of our ethical obligations to the larger communities of life, of which humans are a part. See especially the writing of Aldo Leopold (1947). This philosophy is also referred to as *ethical holism*.
- *Deep Ecology*: The central insight that we must work to overcome the mental and conceptual bifurcation between "nature" and "us." When we reach a point of moral unity, the interests of the natural environment and human society become one (Devall and Sessions, 2001).

In many ways, the key moral question in environmental ethics today is the extent to which (and moral justification for) other forms of life and the natural environment have intrinsic value or inherent moral worth, i.e., a moral value unto themselves irrespective of the value that might be placed on them by and for humans. In considering the ethical and moral dilemmas and judgments before, during, and after a catastrophic event, environmental ethics and values may be an important element, especially when a catastrophic event approaches the most severe level of potential extinction.

3.11 VIRTUE ETHICS AND CATASTROPHIC RESPONSE

The final discussion of ethics explores the potential of another, quite different perspective on ethics known as *virtue ethics*. Virtue ethics emphasizes the character of the individual actor or decision maker. Founded on the ethical thinking of Aristotle (350 BCE) and other Greek philosophers, the notion is that we should focus less on rules of ethical standards for balancing committing interests and rights, and instead concentrate on what the positive character traits and virtues are that modern ethical living requires. In the last decade, especially, there has been considerable interest in returning to virtue-based ethics (Taylor, 1986).

If we are able to nurture or foster virtuous individuals, individuals who will live virtuous lives, then the more principle-based ethics will become less important. Just on an intuitive level, it seems very promising to imagine how cultivating particular virtues—say, sharing, generosity, cooperation—would help immensely in response to a catastrophe where many norms may no longer exist. While not exhaustive, the lists below present some of the more relevant virtues and values associated with the emergency management profession. The final list provides virtues that, in a broader context, might be useful to foster in a resilient culture or society. It is an interesting question to imagine how society might actually go about nurturing and reinforcing these positive virtues. It opens up new and interesting questions about citizenship, for instance, and what is or ought to be required of individuals in the face of serious societal challenges like climate change or a massive hurricane.

BOX 3.10 VIRTUE-BASED ETHICS THAT MAY FOSTER EFFECTIVE EMERGENCY MANAGEMENT

Courage	Cooperation	Acceptance of limitations
Accountability	Curiosity	Care
Compassion	Decency	Benevolence
Frugality	Humility	Wisdom
Fidelity	Patience	Vigilance
Prudence	Respect	Sharing
Reverence	Sacrifice	Simplicity

BOX 3.11 VALUES IN EMERGENCY MANAGEMENT PLANNING AND POLICY

Aesthetics/beauty	Patriotism	Parenting, family values
Personal freedom/liberty	Privacy	
Religious freedom/faith	Public health and safety	Private property rights

BOX 3.12 VIRTUES THAT MIGHT BE USEFUL TO FOSTER IN A RESILIENT CULTURE OR SOCIETY

Attentiveness	Benevolence	Care
Carefulness/cautiousness	Compassion	Cooperation
Courage	Decency	Empathy/sympathy
Forgiveness	Friendliness	Generosity
Honesty/truthfulness	Kindness	Neighborliness
Responsibility/responsiveness	Sense of kinship	Sharing
Sincerity	Trust/confidence	Understanding

A final thought pertains to the issue of *vice* as a counterpart to virtues. Are there negative character or moral traits that work against community and societal resilience and readiness? If so, which and how serious are they? What can be done to mute or discourage these vices in a catastrophic setting?

3.12 NONGOVERNMENTAL ORGANIZATION (NGO) ETHICAL DILEMMAS

Even altruistic organizations, such as the American Red Cross (ARC), face ethical dilemmas. In the mid-1990s, the American Red Cross faced an ethical dilemma concerning where it would and would not assist in managing storm shelters. It created an ethical dilemma when it made the decision not to open shelters in the threatened area until after a hurricane had done its damage. The ARC's stated policy is as follows:

> The Red Cross will not compromise the safety of its shelter residents or its volunteers. That means shelters will not be set up in flood zones or evacuation areas. It spends months planning with state and local governments regarding the location of its shelters (American Red Cross, 2008).

The dilemma the ARC had to face was the choice between its desire to provide shelter from those fleeing the threat of flooding and the requirement that the ARC only operates shelters that are deemed reasonably safe to occupy during a hazard event, such as flooding or a hurricane. Opening a shelter in a flood zone was deemed by the ARC to be inherently unsafe. This decision has had a major impact on evacuation planning in local areas, such as the low-lying Gulf Coast coastal region where hurricane storm surge can penetrate many miles inland. Many state and local officials along the Gulf Coast have expressed great disappointment and frustration with the ARC's decision and the decision continues to harm relations between Gulf Coast states and the ARC. Further frustrating these same officials is that other relief agencies, such as local religious organizations and the Salvation Army, are willing to open and operate shelters in areas where ARC will not.

Following a major disaster or a catastrophe, many NGOs see their mission differently than do government officials. They have been known, at times, to go about their response activities in a manner that can be incongruent with government response operations and other NGOs. The question is: Does an NGO have a duty to do as much good as possible in the manner they believe is right or do they have a duty to comport to government decision making?

3.13 DISCUSSION QUESTIONS

1. Thinking back to the earlier sessions about the types and severity of catastrophic events faced, and the examples of such events given, can you identify an example of an ethical choice, dilemma, or quandary faced by emergency managers or public officials involved in responding to these events?
2. Do you think the type and nature of the ethical dilemmas change according to the type of event of disaster? If so, how?
3. Which ethical quandary or judgment is likely to be the most troubling or difficult? Which decision or policy is most likely to be framed in a nonethical manner (e.g., as primarily an engineering problem, legal issue, or economic dilemma)?
4. Do the ethical responsibilities of emergency managers differ between those who are certified, those who are degreed, and those who are neither?
5. How important is *individual autonomy* in emergency management/catastrophic response decision making?
6. How sacred is autonomy as a value? How and when can this value be trumped by other important values? What restrictions can society/government legitimately place on personal autonomy in the face of a serious public health or natural disaster?
7. How might autonomous choice and freedom be protected in a catastrophic event?

REFERENCES

American Red Cross. 2008. American Red Cross and Gulf Coast residents prepare for Hurricane Ike's arrival. Online at: https://www.crossnet.org/article/0,1072,0_281_8141,00.html (accessed January 10, 2010).

Aristotle. 350 BCE. Nicomachean ethics. Online at: http://classics.mit.edu/Aristotle/nicomachaen.html (accessed January 10, 2010).

Baxter, W. 1974. *People or penguins: The case for optimum pollution.* New York: Columbia University Press.

Beatley, T. 1994. *Ethical land use.* Baltimore, MD: Johns Hopkins University Press.

Devall, B. and G. Sessions. 2001. *Deep ecology.* Layton, Utah: Gibbs Smith Publishers.

Dworkin, G. 1972. Paternalism. *The Monist*, 56: 64–84.

Frankena, W.F. 1973. *Ethics*, Englewood Cliffs, NJ: Prentice-Hall.

Gert, B. 2004. *Common morality: Deciding what to do*, Oxford, U.K.: Oxford University Press.

Josephson Institute on Ethics. (nd). Defining ethics. Online at http://josephsoninstitute.org/ (accessed January 10, 2010).

Kleinig, J. 1983. *Paternalism*. Towota, NJ: Rowman and Allanheld

Leopold, A. 1948. *The Sand County almanac*. Oxford, U.K.: Oxford University Press.

Mill, J.S. 1863. *Utilitarianism*. London: Electric Books.

Mill, J.S. 1859. *On liberty*. London: Longmans, Green and Co.

Partridge, E. (nd). Environmental ethics: An introduction. Online at: http://gadfly.igc.org/e-ethics/Intro-ee.htm (accessed January 10, 2010).

Rawls, J. 1971. *A theory of justice*. Cambridge, MA: Harvard University Press.

Taylor, P. 1986. *Respect for nature*. Princeton, NJ: Princeton University Press.

United Nations. 1948. Universal declaration of human rights. Online at: http://www.un.org/Overview/rights.html (accessed January 10, 2010).

Wilson, E.O. 1975. *Sociobiology, the new synthesis*. Cambridge, MA: Belknat Press.

CHAPTER 4

Political and Legal Issues

4.1 LEARNING OBJECTIVES

As a result of this chapter, the reader should be able to:

- Appraise how the National Response Framework, the National Preparedness Guidelines, the National Incident Management System, and the National Response Plan shape the readiness and response of the United States to extreme events.
- Compare the principles of federalism with the exceptions to federalism (e.g., major exceptions: *posse comitatus*, use of military forces, police roles of military forces, etc.)
- Critique current federal government plans for catastrophe readiness and response and potential federal system breakdowns in potential future catastrophic events.
- Appraise the political structure of the U.S. emergency management system, and the use of the National Guard, Coast Guard, and other military forces in catastrophe response.
- Critique the challenges of interjurisdictional partnership issues.

4.2 KEY TERMS AND PHRASES

- Business Impact Analysis (BIA)
- Business Process Analysis (BPA)
- Continuity of Government (COG)
- Continuity of Operations (COOP)
- Department of Defense, role in a domestic event
- Emergency Management Assistance Compact (EMAC)
- Emergency Support Functions (ESF)

- Federal Continuity Directives (FCD) 1 and 2
- Federal Coordinating Officer (FCO)
- Federal Emergency Management Agency's (FEMA's) Emergency Management Performance Grant (EMPG)
- Federal Preparedness Circular 65 (FPC-65)
- Federalism, principles of and exceptions to
- FEMA's *Comprehensive Preparedness Guide*, CPG 101
- Hazard Mitigation Grant Program (HMGP)
- Incident Management Assistance Team (IMAT)
- Intentional acts that harm others
- Intergovernmental Collaboration
- Joint Field Office (JFO)
- Mission Essential Functions (MEFs) Primary Mission Essential Functions (PMEFs)
- Negligence
- National Essential Functions (NEFs)
- National Priorities of the National Preparedness Guidelines
- National Response Coordination Center (NRCC)
- National Response Framework (NRF), Catastrophic Incident Annex
- Political dimensions of disasters
- Preliminary Damage Assessment (PDA)
- Posse Comitatus Act
- Postdisaster rebuilding of resilient and sustainable communities
- Predisaster recovery plans (both short term and long term)
- Presidential Decision Directive 67 (PDD-67)
- Presidential Directives
- Primary Mission Essential Functions (PMEFs)
- Regional Response Coordination Centers (RRCC)
- Stafford Act; *Emergency Essential Assistance* and *Major Disaster* declarations
- State Coordinating Officer (SCO)
- State Legal Protection Laws for Volunteers
- State public "employee" (including volunteers) indemnification statutes
- Strict liability
- Tort Law
- U.S. Northern Command (USNORTHCOM)
- Vicarious liability
- Windows of opportunity

4.3 OVERVIEW

BOX 4.1 OVERVIEW

Chapter 5 provides an overview of the political and legal framework for planning and response to catastrophic events. Political factors as well as organizational dynamics are included to provide a basis for understanding the complex environment in which planning and responding to extreme events may take place. The literature reflected in political science and public administration provides us with insights into conflicts that arise in highly stressful events and the nature of the problems that evolve from our attempts to deal with catastrophes.

First, we will explore how FEMA's *Comprehensive Preparedness Guide* (CPG 101), the *National Incident Management System* (NIMS), and the federal *National Response Framework* (NRF) shape the readiness and response of the United States to extreme events, such as catastrophes. We assume in this discussion that the reader has learned about these documents previously, but there have been a number of recent changes to the federal emergency response system doctrine, including the replacement of the National Response Plan with the NRF and the replacement of the FEMA's 2007 *National Preparedness Guidelines* with the *Comprehensive Preparedness Guide*, CPG 101.

It is important to emphasize that the basic premise of NIMS and NRF is that initial incident management is the responsibility of local government conducted under the template of NIMS. The NRF builds on NIMS and provides the structure and mechanisms for national-level policy and direction for timely and effective support to state and local government activities. The NRF provides guidelines and procedures to seamlessly integrate federal capabilities and resources into the local response during emergencies and disasters.

4.4 FEMA COMPREHENSIVE PREPAREDNESS GUIDE

The FEMA Comprehensive Preparedness Guide, CPG 101 (online at: http://www.fema.gov/about/divisions/cpg.shtm) provides general guidelines on developing Comprehensive Emergency Operations Plans (EOPs) to state and local emergency managers. Note that throughout this chapter, the term *state and local governments* also includes territorial and tribal governments. It promotes a common understanding of the fundamentals of planning and decision making to help emergency planners examine a hazard and produce integrated, coordinated, and synchronized plans. This guide helps emergency managers in state and local governments in their efforts to develop and maintain a viable all-hazard EOP.

CPG 101 continues the more than 50-year effort to provide guidance about emergency operations planning to state and local governments. Some predecessor material can be traced back to the 1960s-era Federal Civil Defense Guide. Long-time emergency managers also will recognize the influence of Civil Preparedness Guide 1-8, Guide for the Development of State and Local Emergency Operations Plans, State and Local Guide (SLG) 101, Guide for All-Hazards Emergency Operations Planning, in this document. While CPG 101 maintains its link to the past, it also reflects the changed reality of the current emergency planning environment.

BOX 4.2 PURPOSE OF THE COMPREHENSIVE PREPAREDNESS GUIDE

- Organize and synchronize national (including federal, state, local, tribal, and territorial) efforts to strengthen national preparedness.
- Guide national investments in national preparedness.
- Incorporate lessons learned from past disasters into national preparedness priorities.
- Facilitate a capability-based and risk-based investment planning process.
- Establish readiness metrics to measure progress and a system for assessing the nation's overall preparedness capability to respond to major events, especially those involving acts of terrorism.

The guide reinforces the fact that preparedness/planning is a shared responsibility. They were developed through an extensive process that involved more than 1,500 federal, state, and local officials and more than 120 national associations. They also integrate lessons learned following Hurricane Katrina and a 2006 review of states' and major cities' emergency operations and evacuation plans.

The vision of the guide is *"a nation prepared with coordinated capabilities to prevent, protect against, respond to, and recover from all hazards in a way that balances risk with resources and need."* The guide covers a range of readiness initiatives that include:

- All-hazards
- Risk-based
- A call to action

BOX 4.3 ELEMENTS OF THE COMPREHENSIVE PREPAREDNESS GUIDE

- The *National Preparedness Vision* provides a concise statement of the core preparedness goals for the nation.
- The 15 *National Planning Scenarios* identify a set of high-consequence threat scenarios of both potential terrorist attacks and natural disasters.
- The *Universal Task List* (UTL) is a menu of some 1,600 unique tasks that can facilitate efforts to prevent, protect against, respond to, and recover from the major events that are represented by the National Planning Scenarios.
- The *Target Capabilities List* (TCL) defines 37 specific capabilities that communities, the private sector, and all levels of government should collectively possess in order to respond effectively to disasters.

The guide establishes a capabilities-based approach to preparedness. Simply put, a capability provides the means to accomplish a mission. It addresses preparedness/planning for all emergency management mission areas: prevention, protection, response, and recovery. Some capabilities cut across all mission areas and, therefore, are placed in a common mission area.

The guide includes a series of national priorities to guide preparedness efforts that meet the nation's most urgent needs. The priorities reflect major themes and recurring issues identified in national strategies, presidential directives, state and urban area homeland security strategies, the Hurricane Katrina reports, and other lessons-learned reports. FEMA intends to update the priorities over time as it implements the guide or encounters changes in the emergency management strategic environment.

BOX 4.4 FOR AN EXCELLENT ...

For an excellent discussion of the U.S. national response system, see Davis et al. (2007). *Hurricane Katrina lessons for army planning and operations*. Santa Monica, CA: Rand Corporation.

The National Preparedness System provides a way to organize preparedness activities and programs pursuant to the *Comprehensive Preparedness Guide*. The desired end-state of the U.S. National Preparedness System is to achieve and sustain coordinated capabilities to prevent, protect against, respond to, and recover from all hazards in a way that balances risk with resources. Policy and doctrine involves ongoing management and maintenance of national policy and doctrine for operations and preparedness, such as the NIMS, National Response Plan (NRP), Department of Homeland Security (DHS) National Infrastructure Protection Plan (NIPP) and the guidelines. Planning and resource allocation involves application of common planning processes and tools by government officials, working with the private sector, nongovernmental organizations, and individual citizens to identify requirements, allocate resources, and build and maintain coordinated capabilities that are prioritized based on risk. Training, exercises, and lessons learned involve delivery of training and exercises and performance evaluation to identify lessons learned and share effective practices. Assessment and reporting involves assessments based on established readiness metrics and reporting on progress and effectiveness of efforts to achieve the vision of the guidelines.

4.5 THE NATIONAL INCIDENT MANAGEMENT SYSTEM (NIMS)

BOX 4.5 NATIONAL PRIORITIES FOR THE COMPREHENSIVE PREPAREDNESS GUIDE

- Expand regional collaboration
- Implement the NIMS and the NRF
- Implement the NIPP
- Strengthen information sharing and collaboration capabilities
- Strengthen interoperable and operable communications capabilities
- Strengthen chemical, biological, radiological, nuclear, and explosive (CBRNE) detection, response, and decontamination capabilities
- Strengthen medical surge and mass prophylaxis capabilities
- Community preparedness: strengthening planning and citizen capabilities

NIMS provides a consistent nationwide template to enable federal, state, and local governments; the private sector; and nongovernmental organizations to work together effectively and efficiently to prepare for, prevent, respond to, and recover from domestic incidents. It provides a set of standardized organizational structures, such as the Incident Command System (ICS), multiagency coordination systems, and public information systems, as well as requirements for processes, procedures, and systems designed to improve interoperability among jurisdictions and disciplines in various areas.

NIMS has not been tested in a true catastrophe, assuming you, the reader, agree that Hurricane Katrina was not a true catastrophe and there are those in both academia and government who are concerned that the creation of emergent groups in a catastrophe, who may be unable or unwilling to operate within NIMS, will post a major challenge. Emergency managers may be faced with the difficult dilemma in a catastrophe of having to either turn away assistance or allow emergent groups to operate outside of NIMS. This issue is one of the main drivers for the development of hazard-specific catastrophic annexes to the NRF. Whether these NRF annexes can be successfully translated to the NIMS architecture remains to be seen.

BOX 4.6 NIMS STANDARDIZED ORGANIZATIONAL STRUCTURES

- Training
- Resource management
- Personnel qualification and certification
- Equipment certification
- Communications and information management
- Technology support
- Continuous system improvement

The NRP uses the comprehensive framework provided by the NIMS to provide the structure and mechanisms for:

- National-level policy and operational direction for federal support to state and local emergency managers.
- Exercising direct federal authorities and responsibilities as appropriate under the law.

4.6 THE NATIONAL RESPONSE FRAMEWORK (NRF)

The NRF presents the guiding principles that enable all response partners to prepare for and provide a unified national response to emergencies and disasters, from the smallest incident supposedly to the largest catastrophe. This important document establishes a comprehensive, national, all-hazards approach to domestic incident response (for more information, see: http://www.fema.gov/emergency/nrf/mainindex.htm). As we explored in Section I of this book, the question as to whether an "all-hazards" approach will work in a catastrophe remains unresolved. If Hurricane Katrina was a catastrophe, which is debatable, most would agree that the precursor to the NRF, the National Response Plan (NRP), didn't work particularly well, raising numerous questions concerning its utility in a "larger-scale" catastrophe. As we progress through this text, we will revisit this issue several more times.

The NRF defines the key principles, roles, and structures that organize the way we respond as a nation. It describes how communities, tribes, states, the federal government, and private sector and nongovernmental partners apply these principles for a coordinated, effective national response. It also identifies special circumstances where the federal government exercises a larger role, including incidents where federal interests are involved and in catastrophic incidents where a state would require significant support. The NRF enables first responders, decision makers, and supporting entities to provide a unified national response.

In recent years, the United States has faced an unprecedented series of disasters and emergencies, and as a result our national response structures have evolved and improved to meet these threats. The NRF reflects those improvements and replaces the former NRP. Again, it must be noted that these "improvements" include lessons learned from disasters and emergencies. In other words, changes have been in response to the human experience of the emergency management practice. The infrequent occurrence of catastrophes brings limitations on the human experience of emergency managers and government officials. There are those within government and academia who believe that this lack of experience results in a bias of the NRF toward dealing with disasters and emergencies and not catastrophes.

The NRF represents a natural evolution of the national response architecture. Although the NRF was originally called a *plan*, it was actually a framework written to guide the integration of local, state, and federal emergency and disaster response efforts. By adopting the term *framework* within the title, this document is now more accurately aligned with its intended purpose. One significant difference between the NRP and NRF is the latter also focuses on the role of the private sector and nongovernmental organizations (NGOs). The NRF is written for senior elected and appointed leaders, such as federal department or agency heads, governors, mayors, tribal leaders, city or county officials, and leaders of NGOs and the private sector—those who have a responsibility to provide an effective response to preserve the health, safety, and welfare of the community. It informs emergency management practitioners, explaining the operating structures and systems used routinely by first responders and emergency managers at all levels of government. It also is augmented with online access to supporting documents, further training, and an evolving resource for exchanging lessons learned. The reader should be careful not to be left with the impression that the NRF drives local response.

It is important to emphasize that the basic premise of NIMS and the NRF is that initial incident management is the responsibility of local government that is conducted under the template of NIMS. The NRF builds on NIMS and provides the structure and mechanisms for national-level policy and direction for timely and effective support to state and local government activities.

The NRF provides guidelines and procedures to seamlessly integrate federal capabilities and resources, known as Emergency Support Functions (ESFs), into the local response. This is what was referred to in the above observations on the NRF. If the reader is unfamiliar with this concept, he or she is referred to FEMA Independent Study courses IS-700a and 800b online at: http://training.fema.gov/EMIWeb/IS/is700a.asp *and* http://training.fema.gov/EMIWeb/IS/IS800b.asp to better understand NIMS/NRF. In fact, the reader may have taken the old 700 and 800 that were based on the NRP and may not have taken the updated versions (700a and 800b) that are based on the NRF.

The Catastrophic Incident Annex to the NRF (NRF-CIA) establishes the context and overarching strategy for implementing and coordinating an accelerated, proactive national response to a catastrophic incident. The scope of the NRF-CIA is

> A catastrophic incident, as defined by the NRF, is any natural or manmade incident, including terrorism that results in extraordinary levels of mass casualties, damage, or disruption severely affecting the population, infrastructure, environment, economy, national morale, and/or government functions. A catastrophic incident could result in sustained nationwide impacts over a prolonged period of time; almost immediately exceeds resources normally available to state, tribal, local, and private-sector authorities in the impacted area; and significantly interrupts governmental operations and emergency services to such an extent that national security could be threatened. These factors drive the urgency for coordinated national planning to ensure accelerated federal and/or national assistance.
>
> Recognizing that federal and/or national resources are required to augment overwhelmed state, tribal, and local response efforts, the NRF-CIA establishes protocols to pre-identify and rapidly deploy key essential resources (e.g., medical teams, search and rescue teams, transportable shelters, medical and equipment caches, etc.) that are expected to be urgently needed/required to save lives and contain incidents.

> Upon the occurrence of a catastrophic incident, or in advance if determined by the Secretary of Homeland Security, the government will deploy federal resources, organized into incident-specific "packages," in accordance with the NRF-CIS and in coordination with the affected state and incident command structure.
>
> Where state, tribal, or local governments are unable to establish or maintain an effective incident command structure due to catastrophic conditions, the federal government, at the direction of the Secretary of Homeland Security, may establish a unified command structure, led by the Unified Coordination Group (UCG), to save lives, protect property, maintain operation of critical infrastructure/key resources (CIKR), contain the event, and protect national security. The federal government shall transition to its role of coordinating and supporting the state, tribal, or local government when they are capable of reestablishing their incident command. (From Catastrophic Incident Annex CAT-2 Catastrophic Incident Annex, November 2008)

The NRF-CIA is primarily designed to address no-notice or short-notice incidents of catastrophic magnitude, where the need for federal assistance is obvious and immediate, where anticipatory planning and resource prepositioning are precluded, and where the exact nature of needed resources and assets is not known in advance. Appropriately tailored assets and responses identified in the NRF-CIA, as well as other select federal resources and assets, may also be deployed in support of a projected catastrophic event (e.g., a major hurricane) with advance warning in support of the anticipated requests of state and local governments. (For more information on the NRF-CIA, go to: http://www.fema.gov/pdf/emergency/nrf/nrf_CatastrophicIncidentAnnex.pdf.)

4.7 PRINCIPLES OF FEDERALISM AND EXCEPTIONS TO FEDERALISM

The federal system in the United States is one that is based on enumerated powers specifically granted to it in the U.S. Constitution and statutes. The Tenth Amendment to the Constitution provides that powers not delegated to the federal government nor prohibited by it to the states are reserved to the states. Key federal powers include the collection of taxes and duties, payment of debts, and providing for welfare and the common defense. Other federal powers include regulating commerce among multiple states and foreign nations, establishing a militia, and protecting civil rights and liberties.

Faber and Chen (2006) notes a weakness in the federal system may have been exposed by Hurricane Katrina. Because of our federalized system, as in the Katrina response, FEMA and the DHS are not in complete control during and responding to a future catastrophe. As a result of this shared, layered, power arrangement, there is no single authority within the United States to effectively prepare, respond, mitigate, and aid in recovery from catastrophes.

The question is often asked: Why can't federal authorities simply respond to disasters and catastrophes without waiting for a formal request from the state governor? There are a number of reasons, including:

- The U.S. Constitution spells out the powers and responsibilities of the federal government and reserves all other powers and responsibilities for the individual states. The federal government has only limited powers to respond without the state's consent, such as a civil uprising or an attack upon the United States.
- Local governments have only the powers and responsibilities delegated by their state governments and state constitutions and are often restricted in their authority to regulate land-use, building standards, and other functions and in their authority to raise taxes, provide services, and exercise discretion in policymaking.

As the reader can see, primary authority and responsibility for responding to emergencies, disasters, and catastrophes resides in the states, but the federal government may assume authority and responsibility for an event that rises to the level that may threaten national security or the very existence of the government. As a result, federal agencies can act directly and assume leadership in crisis situations related to war, civil uprising, terrorism, and certain types of criminal activity, such as kidnappings and bank robberies that are presumed to be interstate crimes. In the event federal authorities may have the "lead" in these circumstances, local and state entities are still responsible for coordinating the emergency management response. Put another way, certain federal agencies, such as the Department of Justice (specifically, the FBI) are tasked by federal law to take the lead in criminal investigations in the examples listed previously, while local government still has incident/emergency management responsibility.

Think about how local residents and officials might respond if state or federal officials arrived in response to a disaster, set up a command center, and simply took over local emergency response operations. As we look at catastrophic scenarios, one could ask under what circumstances such a scenario might be reasonable or justifiable. Would the incapacitation of local and state authorities in a catastrophe warrant such a federal role? Or would the refusal of a local authority to respond or, more interesting, an ineffective response, warrant such a federal role? Conversely, under what circumstances might federal intervention make it very inappropriate and possibly even more difficult to respond effectively to the needs of victims?

The Stafford Act provides for two distinct circumstances under which the president may deploy federal troops. They are:

1. *Emergency Essential Assistance*: The Stafford Act requires that the governor must determine the situation to be of great severity such that the state is unable to manage the situation without federal help. The governor must execute the state response plan and activate the state National Guard units under his control before the president may activate and deploy either National Guard units from other states, in-state and out-of-state military reserve units (which are not under the governor's control), or regular military units.
2. *Major Disaster*: This provision is similar to the above criteria, but mandates the specification of what is needed and what is not required. The governor must explain what is currently being utilized at the state and local level. Note that the governor may keep the National Guard units under state control when he requests additional military support from the president.

The obvious question is what happens in a catastrophe where the governor is incapacitated or is unable to communicate his/her state's requirements to the federal government? Does the president have the authority to act without the consent of the governor? The determination of whether the president can send federal troops in response to a catastrophe without the governor's concurrence is rooted in federal law. The Posse Comitatus Act (18 U.S.C. 1385), along with other related laws and administrative provisions, prohibits the use of the military to execute civilian laws unless expressly authorized by the Constitution or an act of Congress. The Posse Comitatus Act (18 U.S.C. Section 1835) provides that: "Whoever, except in cases and under circumstances expressly authorized by the Constitution or Act of Congress, willfully uses any part of the Army, Navy, Marines, or the Air Force as a Posse Comitatus or otherwise to execute the laws shall be fined under this title or imprisoned not more than two years, or both." Note that the Coast Guard is not included in this restriction, giving the president an important response tool that we all saw work effectively in the Hurricane Katrina response.

Note that Congress can provide for means of allowing the military to assist in response to a catastrophe. This, of course, assumes that Congress exists, as a body, during and after a catastrophe. The Stafford Act is a key means of allowing the president to use the military resources in a catastrophe. Congress has made a number of exceptions to the act, which permits military involvement in law enforcement. For example, Congress has enacted a number of statutes that authorize the president to use military forces to suppress insurrections and civil uprisings and disturbances (10 U.S.C. 331-335). If these statutes were to be invoked, the president could use active or reserve components of the military to put down a rebellion or to control a civil uprising and disturbance. Another important exception relates to the Coast Guard, as pointed out earlier, which Congress has vested with broad domestic law enforcement authority.

Under these statutory provisions, the Coast Guard and Coast Guard Reserve can participate in the enforcement of maritime, customs, and certain other federal laws. (For more information on the Posse Comitatus Act, see Congressional Research Service (CRS) Report RS20590, *The Posse Comitatus Act and Related Matters: A Sketch*, and CRS Report RS22266, *The Use of Federal Troops for Disaster Assistance: Some Legal Issues.*)

Within the statutory exceptions to the Posse Comitatus Act involving the use of military forces, there are a number of "police" roles that military forces can play. Department of Defense regulations provide emergency powers that authorize actions to be taken under the inherent right of the U.S. government to ensure the preservation of public order and carry out governmental operations (Department of Defense (DoD) Cooperation with Civilian Law Enforcement Officials, DoD Directive 5525.5— Encl. 4— section E4.1.2.3). Federal use of the military may be taken to restore government functioning and public order when sudden and unexpected civil disturbances, disaster, or calamities (catastrophes) seriously endanger life and property and disrupt normal governmental functions (includes civic disturbances, such as earthquakes, fires, floods, or other such calamity that endangers life). The DoD may also provide support to local authorities for rescue, evacuation for/of emergency medical services, debris removal, and restoration of essential services and technical assistance in areas such as critical infrastructure restoration.

In 2008, Congress passed legislation entitled Title 6. *Domestic Security Chapter 1. Homeland Security Organization Coordination with Non-Federal Entities* (2008), 6 USC § 466, whose purpose was to reaffirm congressional support for the applicability of posse comitatus in the "post 9/11" era.

4.8 INTERGOVERNMENTAL COLLABORATION: A CORNERSTONE OF EFFECTIVE CATASTROPHIC PLANNING AND RESPONSE

Effective catastrophic planning requires regional collaborations between countries, all levels of government, NGOs, and the private sector. This section focuses on the integration of state and federal agency response efforts. This collaboration is not easy to achieve as jurisdictions often have competing interests. Catastrophes by definition overwhelm state and local governments and often national response capacities. Under this view, it is critical that preparedness/planning efforts should be regional in nature and include multistate collaborations. In addition, these efforts should involve the numerous federal agencies that have regulatory or monitoring responsibilities. Unfortunately, such a regional approach is not mandated by federal statute, leaving the state and local jurisdictions to participate at their own discretion and peril. It is worth noting that interstate evacuation planning activities by Louisiana were resisted by neighboring states for years until Hurricane Katrina struck in 2005, forcing these states to accept responsibility for cooperating. In the end, the evacuation executed for Hurricane Katrina was a multistate collaboration, though not necessarily among totally willing states. One of the key challenges in catastrophic planning and response is the tendency of organizations to maintain their independence and autonomy while a broader community-wide focus and interorganizational interdependence is needed. (Tierney et al., 2001.)

Another interesting dynamic resulting from the Hurricane Katrina experience is that states that are being asked to host evacuees want to be made financially whole as a condition of accepting evacuees. Following Hurricane Katrina, "host states" had to request reimbursement through Louisiana. This reimbursement took many months to receive and, in some cases, host states gave up and absorbed the costs. This issue is one that continues to be an ongoing subject of discussion between states and FEMA and, at the time this book was written, FEMA was in the final steps of revising the Stafford Act regulations to allow FEMA to directly reimburse host states.

In recent years, the importance of interstate mutual aid agreements, known as EMACs (emergency management assistance compacts), have grown in importance. Only a governor can enter into such an agreement on behalf of his/her state. EMACs represent one of the few sustained and successful examples of interstate cooperation that may be critical in responding to a catastrophe. EMAC agreements contain reimbursement clauses that are recognized by FEMA, which means that, if there is a declared emergency of major disaster and another state assists the impacted state, FEMA will reimburse the assisting state for its costs. A governor has the authority to request EMAC emergency assistance.

Problems and challenges in catastrophic preparedness initiatives are often seen in organizational decision making. In a normal noncrisis time, people and organizations typically try to make rational choices. However, as we have already learned, in a catastrophic situation, a rational model is more difficult and achieving it is often convoluted, even if at all. Because of the constraints in a catastrophe, rational approaches rarely are achieved. Interorganizational conflict and collaboration clearly have an impact on developing sound public policy in catastrophic planning and response. Burby and May (1997, p. 264) note that there are five principles that offer a foundation for sound public policy and avoid the disaster cycle of recurring disasters in vulnerable locations. These principles appear to fit well with how government can mitigate the severity of catastrophes as well:

1. Governments must limit the practice of subsidizing the risks involved in using hazardous areas.
2. Governments must build and share a base of knowledge about the nature of risks and sustainable ways of living with hazards.
3. All levels of government must develop commitment and capacity to change the way they manage the use of hazardous areas.
4. Governments must do a better job of coordinating policies to manage exposure to hazards with policies to accomplish economic, social, and environmental objectives through community development.
5. Governments must foster innovation in governance and land management to better match institutional systems and tools with the nature of the problems posed by natural hazards.

BOX 4.7 6 USC § 466: SENSE OF CONGRESS REAFFIRMING THE CONTINUED IMPORTANCE AND APPLICABILITY OF THE POSSE COMITATUS ACT

(a) Findings. Congress finds the following:
   (1) Section 1385 of title 18, United States Code (commonly known as the "Posse Comitatus Act"), prohibits the use of the Armed Forces as a posse comitatus to execute the laws except in cases and under circumstances expressly authorized by the Constitution or Act of Congress.
   (2) Enacted in 1878, the Posse Comitatus Act was expressly intended to prevent United States Marshals, on their own initiative, from calling on the Army for assistance in enforcing Federal law.
   (3) The Posse Comitatus Act has served the Nation well in limiting the use of the Armed Forces to enforce the law.

(continued on next page)

(4) Nevertheless, by its express terms, the Posse Comitatus Act is not a complete barrier to the use of the Armed Forces for a range of domestic purposes, including law enforcement functions, when the use of the Armed Forces is authorized by Act of Congress or the President determines that the use of the Armed Forces is required to fulfill the President's obligations under the Constitution to respond promptly in time of war, insurrection, or other serious emergency.

(5) Existing laws, including chapter 15 of title 10, United States Code [10 USCS §§ 331 et seq.] (Commonly known as the "**Insurrection Act**"), and the Robert T. Stafford Disaster Relief and Emergency Assistance Act (42 U.S.C. 5121 et seq.), grant the President broad powers that may be invoked in the event of domestic emergencies, including an attack against the Nation using weapons of mass destruction, and these laws specifically authorize the President to use the Armed Forces to help restore public order.

Public agencies at the local, state, and national levels must collaborate both vertically and horizontally to attain effective catastrophic planning and response efforts. Agencies at all levels attempted to integrate their plans prior to Hurricane Katrina, but many of the postevent assessments found that their planning and response efforts were not effective. Katrina demonstrated that blaming other organizations for operational problems was a common dynamic. Through proper planning, all levels of government can reduce or eliminate this phenomenon that results from the chaos that may be present in a response to a catastrophe. Some of the factors that must be addressed to ensure that catastrophic preparedness/planning efforts reflect collaboration among and between all levels of government include:

- Catastrophes are political in nature and involve interorganizational conflict and blame.
- Conflict between organizations, to a significant degree, is part of response operations. Disagreements may evolve concerning some of the following issues:
  - Who will be given authority over incident command?
  - Which organizations will be assigned to what they perceive as menial or less visible tasks?

    - Which organizations will be given additional resources and responsibilities as the nature of the event unfolds?
    - Who will get credit or blame for the outcome of the event response?
- Because of this conflict, some organizations restrict communication and coordination with others simply because interorganizational rivalries exist.
- Politics are also prevalent when disasters are declared (and will also be present in a catastrophe) and when resources begin to be distributed. (Political issues will be a large source of these discrepancies.)
    - There will be disagreement about who should get help first.
    - Favoring major supporters or ignoring those who have significantly less power may occur following a catastrophe.
    - Interorganizational communication, collaboration, and control will be essential in an effective response, but will it occur? (NIMS has been a critical initiative in order to address these concerns, but remains untested in a true catastrophe and after-act reports indicate it didn't work well in the Hurricane Katrina response.)

Some strategies that emergency managers can undertake to reduce interorganizational conflict in planning for and responding to a catastrophe include:

- Get to know departmental leaders in the jurisdiction and those of all other organizations that offer potential assistance or become a player.
- Try to develop a rapport with them and amongst each other.
- Plan together and clarify responsibilities before a catastrophe strikes.
- Promote the use of unified command in order to prioritize objectives across disciplines and/or jurisdictions, ensuring the elected leader understands what a written delegation of authority is and how and when to use it.
- It is much easier to find consensus during preparedness/planning rather than resolve disagreements after response operations have begun.
- Reason with organizations, point out the "greater good" mission that will be essential for a successful catastrophic response.
- Show the merit of cooperation, communication, and coordination (e.g., how this will help disaster victims and speed up response and recovery operations).

- Go to the political figures for assistance, enlist their support and buy-in for the mission.
- If all else fails, ask political leaders to settle differences or enforce decisions.

There is no doubt that decision making is difficult in a catastrophe and there are numerous reasons why this occurs. Dror (1988) identified reasons why decision making under catastrophic conditions and strategies to address them is difficult:

- Catastrophes involve adversity and are characterized by injury, death, and destruction and demand the immediate attention of decision makers.
- Time is critical in a catastrophe and because people's lives and well-being are at stake, there is incredible pressure for decision makers to act quickly and even prematurely.
- There are no easy decisions in catastrophes for they are often accompanied by situations where there are drawbacks to nearly every decision that needs to be made.
- Decision making during response operations is challenging, as uncertainty is an expected correlate of a catastrophe.
- The physical and emotional demands placed on decision makers are great and they may impair effective decision making. (This will be discussed further in Section V of this book.)

Dror (1988) also offers strategies for effective decision making in a disaster response. He recommends designing preferable models. This strategy is similar to a rational model as it entails studying the situation or problem in detail, determining the gap that exists between the goal and reality, and intervening to adapt the process to the desired outcome. Dror also recommends *debugging*, which involves an acute observation of the decision process in order to correct potential weaknesses and mistakes as they become apparent and, lastly, think critically; people often fail to "think outside the box" in a catastrophe.

4.9 FEDERAL GOVERNMENT PLANS FOR CATASTROPHE READINESS AND RESPONSE

The primary federal system for responding to a catastrophe is the Robert T. Stafford Disaster Relief and Emergency Assistance Act, 42 U.S.C. Section 5170. The Stafford Act provides an orderly and continuing means of assistance by the federal government to state and local governments in carrying out their responsibilities to alleviate the suffering and damage that result from an emergency, disaster, or catastrophe. It supplements state and local resources in declared emergencies or major disasters where state and local resources have been overwhelmed. Except to the extent that an emergency involves primary federal interests, the declaration of a major disaster must be triggered by a request to the president from the governor. The president, in response to a governor's request, may declare an "emergency" or "major disaster" in order to provide federal assistance under the Act (44 CFR 206.2). The Stafford Act supports state and local government response efforts. Federal operational components that may be activated under authorization of the Stafford Act include the National Response Coordination Center (NRCC) at FEMA Headquarters in Washington, D.C., FEMA Regional Response Coordination Centers (RRCC) located at FEMA's 10 regional offices, Joint Field Office (JFO) (operated jointly with the state) that will be located at a location within the declared state that is mutually agreeable to both the state and FEMA, and Disaster Recovery Centers (DRCs).

Within the NRCC, RRCCs, and the JFOs, Emergency Support Functions (ESF) provides the structure for coordinating federal interagency support for a federal response to an incident with that of the state. They are mechanisms for grouping functions most frequently used to provide support to states and federal-to-federal support, both for declared major disasters and emergencies under the Stafford Act and for non-Stafford Act incidents. States organize their response along the same ESF functions as the federal government and work closely together in the JFO. While ESFs are typically assigned to a specific section at the NRCC or in the JFO/RRCC for management purposes, resources may be assigned anywhere within the unified coordination structure. Regardless of the section in which an ESF may reside, that entity works in conjunction with other JFO sections to ensure that appropriate planning and execution of missions occur. The ESF structure will be discussed further in Section IV of this book.

The DHS National Operations Center and FEMA's NRCC continually monitor potential major disasters and emergencies. When advance warning is received, FEMA may deploy—and may request that other federal agencies deploy—liaison officers and personnel to a state emergency operations center to assess the emerging situation. FEMA regional offices may open one or more RRCCs and they may be fully or partially activated. Facilities, such as mobilization centers, may be established to accommodate federal personnel, equipment, and supplies. Since Hurricane Katrina, FEMA has continued to develop another field deployable asset called an Incident Management Assistant Team (IMAT). IMATs are intended to be forward advance elements of a JFO and include the president's federal coordinating officer for that specific disaster.

Immediately after a major, damaging event, local emergency personnel respond and assess the situation. If necessary, these officials seek additional resources through mutual aid and assistance agreements and through the state. State officials also review the situation, mobilize state resources, use interstate mutual aid and assistance processes (such as the EMAC to augment state resources), and provide situation assessments to the RRCC at the FEMA regional office. The governor activates the state emergency operations plan, declares a state of emergency, and may request a state/FEMA joint Preliminary Damage Assessment (PDA). The state and federal officials conduct the PDA in coordination with local officials as required and determine whether the impact of the event warrants a request for a presidential declaration of a major disaster or emergency. Based on the results of the PDA, the governor may request a presidential declaration specifying the kind of federal assistance needed.

After a major disaster or emergency declaration, an RRCC coordinates initial regional and field activities until a JFO is established. If there was advance warning, an IMAT may have been deployed and the RRCC will coordinate with the state through the IMAT. Regional teams assess the impact of the event, gauge immediate state needs, and make preliminary arrangements to set up field facilities. If regional resources are or may be overwhelmed or if it appears that the event may result in particularly significant consequences, depending on the scope and impact of the event, the NRCC carries out initial activations and mission assignments and supports the RRCC.

The governor appoints a state coordinating officer (SCO) to oversee state response and recovery efforts. The federal coordinating officer (FCO), appointed by the president in a Stafford Act declaration, coordinates federal activities in support of the state. A JFO may be established locally to provide a central point for federal, state, and local executives to coordinate their support to the incident. The unified coordination group leads the JFO and typically consists of the FCO, SCO, and senior officials from other entities with primary statutory or jurisdictional responsibility and significant operational responsibility for an aspect of an incident. This group meets to develop a common set of objectives and a coordinated initial JFO action plan. The group coordinates field operations from a JFO. In coordination with state and local agencies, Emergency Support Functions assess the situation and identify requirements. Federal agencies provide resources under FEMA mission assignments or their own authorities.

As immediate response priorities are met, recovery activities begin. Federal and state agencies assisting with recovery, reconstruction, and mitigation activities convene to discuss needs. The Stafford Act's Public Assistance program provides disaster assistance to states, tribes, local governments, and certain private nonprofit organizations. FEMA, in conjunction with the state, conducts briefings to inform government entities in the impacted area (known as "subgrantees," as the state or tribe will be the grantee) of the assistance that is available and how to apply. Throughout response and recovery operations, FEMA Hazard Mitigation program staff at the JFO looks for opportunities to maximize mitigation efforts in accordance with state hazard mitigation plans that can reduce the impacts of a future damaging event.

As the need for full-time interagency coordination at the JFO decreases, the unified coordination group plans for selective release of federal resources, demobilization, and closeout. Federal agencies work directly with disaster assistance grantees (i.e., state or tribal governments) from their regional or headquarters offices to administer and monitor individual recovery programs, support, and technical services.

State homeland security and emergency management statutes may include a provision that an all-hazard emergency response plan be developed. Such an act should clearly provide for the authority of the governor as well as what local governments may do in a disaster. The provision of the state act also provides that a state emergency management agency and its operational procedures be established.

This discussion is clearly predicated on the continued viability of state and local governments during and following a catastrophe. While it is reasonable to assume that these political entities will remain viable during an emergency and probably during a disaster, a similar assumption during and following a rapid onset catastrophe would be dangerous. That is the underlying purpose of the hazard-specific catastrophe annexes to the NRF.

4.10 ENSURING ENDURING FEDERAL AND STATE CONSTITUTIONAL GOVERNMENTS

Following a catastrophe, segments of federal, state, and local governments as well as NGOs and the private sector may be severely compromised and no longer viable entities. The federal government and its national partners must be prepared to fill potential gaps to ensure continuity of government and public-sector life-sustaining operations and systems. The following discussion describes how the federal government plans to manage such an event and how the federal government is supporting similar efforts at the state level.

Continuity of operations (COOP) and continuity of government (COG) planning have been a part of government operations since at least the Cold War when President Dwight D. Eisenhower provided, via Executive Order, various measures designed to ensure that the U.S. government would be able to continue operating after a nuclear attack. The

COG portion of this effort, intended to preserve our constitution government, remains a national security classified program under the assumption that knowledge of these plans would enable our enemies to potentially disrupt our enduring constitutional government. In addition, these plans remain classified to prevent uproar among the American public, who proponents fear, might panic with the revelation that the government was planning for its own survival in a terrifying postnuclear attack environment, but not the survival of the general public. With the demise of the Soviet empire, both COOP and GOG planning lost their prominence both in government and in the public consciousness. Both efforts have seen a return to relevance in the 2000s. After the September 11, 2001, attacks, many speculated that terrorists might attempt to destroy a large part of the central government and send the country into chaos.

Both COOP and COG plans were fully activated for the first time in response to the September 11 attacks. Since that date, DHS has conducted three exercises to test continuity plans. The first, called "Forward Challenge '04", took place from May 12 and 13, 2004, and included more than 40 government agencies. The second major exercise took place from June 20 to 24, 2005. Titled "Pinnacle," this exercise tested responses to various emergencies, including a hypothetical act of terrorism. "Forward Challenge '06" was the third major exercise, and took place on June 19, 2006, involving nearly 4,000 government personnel. Future similar exercises are in the planning at the time of this writing.

4.10.1 Federal Continuity of Operations (COOP) Planning

The Federal COOP initiative is an unclassified effort within individual federal departments and agencies to ensure continuity of their essential functions across a wide range of emergencies and other events. Americans have come to rely on the government for many essential services, from airport security and air traffic control, to Social Security, veteran's benefits, and Medicare payment processing. While COOP can be activated for an event as small as a threat to an area as small as one locality or federal facility, COOP will play an important role during a catastrophe because it will impact the ability of federal workers to report to their duty stations or alternate duty stations and to continue to deliver essential government services.

To ensure that federal agencies are able to continue their essential functions, President Clinton signed Presidential Decision Directive 67 (PDD-67) requiring all federal executive branch departments and agencies to develop COOP plans. PDD-67 required all federal executive branch departments and agencies to develop plans in response to all hazards and a full spectrum of threats, including:

- Natural
- Manmade (terrorism)
- Technological
- National security emergencies

Today's threat environment makes COOP planning even more important. DHS is the executive department responsible for overseeing federal executive branch-wide COOP planning and, within DHS, FEMA has been designated the executive agency. FEMA has issued Federal Preparedness Circular 65 (FPC-65), which incorporates the topics addressed in previous PDDs and includes more specific and detailed guidance regarding COOP capabilities each federal department and agency must put in place. Federal COOP planning includes the activities of individual departments and agencies and their subcomponents to ensure that their essential functions are performed. Governments at all levels have a fundamental responsibility to provide uninterrupted essential services to the public, regardless of circumstances. COOP planning must incorporate a wide range of emergencies and events, whether natural, manmade, or technological in nature.

BOX 4.8 OBJECTIVES OF COOP PLANNING

- Ensure the performance of an agency's essential functions during a COOP event.
- Reduce loss of life by minimizing damage and losses.
- Ensure the successful succession to office in the event a disruption renders agency leadership unavailable to perform their responsibilities.
- Reduce or mitigate disruptions to operations.
- Ensure that agencies have alternate facilities from which to operate.
- Protect essential facilities, equipment, vital records, and other assets.
- Achieve a timely and orderly recovery from a COOP situation.
- Achieve a timely and orderly reconstitution from an emergency and resume full service to internal and external customers.

COOP activities include:

- Plans and procedures to ensure that essential functions are performed. Plans and procedures may include delegations of authority, orders of succession, notification procedures, and check-in procedures at the alternate facility. Other planning elements include alternate information technology systems and the ability to access critical information, via information technology systems, from an alternative worksite.
- Tests, training, and exercises essential for ensuring a viable COOP capability. Tests, training, and exercises (TT&E) should cover all COOP plans and procedures.

Federal Continuity Directive (FCD) 1, issued in February 2008 (Figure 4.1), provides direction to the federal executive branch for developing continuity plans and programs. Continuity planning facilitates the performance of executive branch essential functions during all-hazards emergencies or other situations that may disrupt normal operations. The ultimate goal of continuity in the executive branch is the continuation of National Essential Functions (NEFs). (For more information on the FCD 1, go to: http://www.fema.gov/pdf/about/offices/fcd1.pdf.)

Federal Continuity Directive 1 (FCD 1)
Federal Executive Branch National Continuity Program and Requirements

February 2008

Homeland Security

FIGURE 4.1 Cover of Federal Continuity Directive (FCD 1), January 2008. (U.S. Department of Homeland Security.)

The provisions of FCD 2, also issued in February 2008 (Figure 4.2), are applicable to the executive departments enumerated in 5 U.S.C. § 101, including the Department of Homeland Security, independent establishments as defined by 5 U.S.C. § 104(1), government corporations as defined by 5 U.S.C. § 103(1), and the United States Postal Service. The departments, agencies, and independent organizations are hereinafter referred to as *agencies*. The continuity program elements outlined herein are for use at all levels of federal executive branch organizations regardless of their location, and are also useful reference documents for nonfederal government and private sector entities.

FIGURE 4.2 Cover of Federal Continuity Directive (FCD 2), January 2008. (U.S. Department of Homeland Security.)

FCD 2 implements the requirements of FCD 1, ANNEX C, and provides guidance and direction to federal executive branch departments and agencies on how to identify their Mission Essential Functions (MEFs) and potential Primary Mission Essential Functions (PMEFs). It includes guidance and checklists to assist departments and agencies in assessing their essential functions through a risk management process and in identifying potential PMEFs that support the NEFs—the most critical functions necessary to lead and sustain the nation during a catastrophic emergency. The FCD 1 and 2 provide direction on the formalized process for submission of a department or agency's potential PMEFs that are supportive of the NEFs. It also includes guidance on the processes for conducting a Business Process Analysis (BPA) and Business Impact Analysis (BIA) for each of the potential PMEFs that assist in identifying essential function relationships and interdependencies, time sensitivities, threat and vulnerability analyses, and mitigation strategies that impact and support the PMEFs. (For more information on the FCD 2, go to: http://www.fema.gov/pdf/about/offices/fcd2.pdf. To learn more about the COOP program, see. http://www.fema.gov/government/coop/index.shtm#2.)

4.10.2 Federal Continuity of Government (COG)

COG is the principle of establishing defined procedures that allow a government to continue its constitutional functions in case of nuclear attack or other catastrophic event. The federal government's COG program is classified and little official information is available to the public. The concept of COG was developed by the British government during World War II to counter the threat of Luftwaffe bombing during the Battle of Britain. The need for COG plans gained new urgency during the Cold War, but waned after the fall of the Soviet Union, only to be revived in the post-9/11 era.

COG planning measures include construction of underground facilities intended to protect a small group of government workers that would continue to carry out our constitutional government and other critical functions to ensure that our form of government endures. Other provisions of the plans include executive orders that designate certain government officials to assume cabinet and other executive branch positions if the primary office holders were killed, injured, or otherwise unable to carry out their duties. There has been a formal line of succession to the presidency since 1792 (currently found in the Presidential Succession Act of 1947, 3 U.S.C. § 19). This runs from the vice president to the speaker of the House of Representatives, president *pro tempore* of the Senate, and then through the cabinet secretaries in a sequence specified by Congress. Most who follow the government's COG planning believe that COG programs were activated in response to the 9/11 attacks. In March 2006, the federal government acknowledged that COG activities did occur on 9/11, but would not further explain what actions were taken and whether they were effective.

4.10.3 State Level COOP/COG Planning

COOP and COG planning is the fundamental responsibility of every government agency that performs an essential function at the state and local level. In order to conduct necessary emergency operations, recovery actions, and other key essential functions during a large-scale or catastrophic event, the agency must have effective COOP plans in place to support continued operations of mission critical functions. COOP efforts also provide the foundational basis for COG programs, such as succession planning, which are designed to ensure the survival of not only leadership at the state and local level, but also an enduring constitutional government. State and local plans to address COOP/COG issues should be consistent with Homeland Security Presidential Directive (HSPD)-20, National Continuity Policy, which provides guidance for state and local governments, as well as private sector organizations to ensure a comprehensive and integrated national continuity program, and FEMA COOP planning guidance. Issues to address include, but are not limited to:

- Delineate essential functions and activities, agency interdependencies, and the resources needed to perform them.
- Establish orders of succession and delegations of authority to key agency positions and establish and maintain current roster(s) of fully equipped and trained COOP personnel with the authority to perform essential functions.

- Provide for the identification and preparation of alternate operating facilities for relocated operations.
- Provide for the regular training, testing, and exercising of COOP personnel, systems, and facilities.
- Provide for reconstitution of agency capabilities, and transition from continuity operations back to normal operations.

State COOP and COG planning is eligible for funding under FEMA's Emergency Management Performance Grant (EMPG) program that supports an all-hazards approach to emergency response, including the development of a comprehensive program of planning, training, and exercises, that provides the foundation for an effective and consistent response to any threatened or actual emergency or major disaster regardless of the cause. As appropriated by the Consolidated Security, Disaster Assistance, and Continuing Appropriations Act, 2009 (Public Law 110-329), the EMPG program provides assistance to state and local governments to enhance and sustain all-hazards emergency management capabilities. States have the opportunity to use EMPG funds to further strengthen their ability to support emergency management mission areas while simultaneously addressing issues of national concern as identified in the "National Priorities" of the *National Preparedness Guidelines*. (To learn more about the EMPG program, go to: http://www.fema.gov/emergency/empg/empg.shtm.)

4.11 LOCAL GOVERNMENT COMPREHENSIVE EMERGENCY RESPONSE PLANS

Local government units are required under many state homeland security and emergency management statutes to develop an all-hazards emergency plan. These are state laws that impose a statutory duty to plan for disasters. The statute may provide that the local government will forward the plan to the state emergency management agency.

Separate from the Stafford Act provisions, Emergency Planning and Community Right to Know legislation adopted at the state level requires that communities plan for extremely hazardous substances, a set of hazardous substances defined by U.S. Environmental Protection Agency (EPA) regulations. Local jurisdictions are required to obtain inventory information from local chemical processors who store, process, use, or transport hazardous substances. If they have any extremely hazardous substances in their inventories, they are required to complete a plan.

Local planning requirements also are associated with hazard mitigation planning. Rather than requiring mitigation planning in state emergency management statutes, many of these planning requirements are associated with qualifying and obtaining mitigation funds from federal and state hazard mitigation programs. Under FEMA's Hazard Mitigation Grant Program, before the state emergency management agency will provide funding for local hazard mitigation projects, the local entity must prepare a hazard mitigation plan. If the local entity fails to complete this plan, they do not qualify for consideration of federal postdisaster mitigation funds.

It should be noted that statutory requirements to develop a local emergency response plan establish a legal duty "to plan" on the part of the local entity. Failure to comply with this statutory duty could lead to a claim of negligence by those harmed by a lack of planning against those local entities that failed to comply with state of local disaster planning statutes and regulations.

4.12 THE POLITICAL STRUCTURE OF THE U.S. EMERGENCY MANAGEMENT SYSTEM

4.12.1 Local Government

Local governments, whether cities, towns, boroughs, villages, counties, or a parish, are central organizations in emergency management since local government has the primary responsibility for public safety, including emergency response following an event. The local chief executive official is usually the person in charge, unless another official has been designated through ordinance or legislation. In general terms, local laws define, with widely varying specificity and scope, who will do what in preparing for, mitigating, responding to, or recovering from emergencies or disasters. The objective here is to establish a legal authority for the development and maintenance of an emergency management program and organization and to define the emergency powers, authorities, and responsibilities of the chief executive official and the emergency manager. Obviously, local legislation must be in conformance with state legislation.

With respect to catastrophic response, states must strongly consider that local governments will not be able to function effectively for extended periods of time. This loss in local government capability can result from the death, incapacitation, or absence of local leadership, elected, appointed, and professional personnel, and first responder personnel.

4.12.2 State Government

States, tribes, as well as U.S. possessions and territories, play many roles and have numerous responsibilities when it comes to emergency management. States enact emergency management legislation, codes, regulations, and fund emergency management activities. States ensure that federal laws are enforced and serve as an interface with federal agencies. They also develop integrated and comprehensive programs across all phases of emergency management, preparedness, response, recovery, and mitigation, and state emergency management agencies coordinate state agency emergency management activities. Another very important role: Just as the federal government assists the states, states assist local governments with both resources and funding.

The role of the governor is vital to emergency management as he/she establishes the state level responsibility for emergency management through the executive branch. The governor declares state emergencies and disasters and directs his/her state's disaster response. The governor also commands state National Guard units that represent a significant response resource resident within each state. As discussed previously, the emerging importance of interstate mutual aid agreements, known as EMACs, have grown in importance.

Governors also possess various emergency powers that vary from state to state. Typical gubernatorial emergency powers include:

- Authority to declare emergencies and disasters
- Suspend state laws
- Mobilize the National Guard within the state
- Seize personal property in direct furtherance of the protection of the health, safety, and welfare of citizens of the state
- Direct evacuations
- Restrict access to damage areas
- Authorize emergency funding

States have various responsibilities in responding to a crisis, including a potential catastrophe. Local authorities and individuals request help from private relief organizations and the state government, which, if warranted, activates its disaster response plan. Generally, governors have, or are granted, the power to use all available state resources needed to respond effectively and efficiently. In many states, governors can suspend state laws or local ordinances if it is determined that the law in question will restrict or prohibit efforts to relieve human suffering caused by the event.

In some states, after a state emergency declaration, the governor may establish economic controls over such resources and services as food, wages, clothing, and shelter in affected areas. Under a state emergency declaration, governors often may issue emergency orders to protect the public from such adverse actions as "price gouging" by unscrupulous merchants. Governors are empowered to mobilize the National Guard units within their state and direct their efforts. Most are also empowered to direct citizen evacuation, to order the control of movement into or out of disaster-declared areas, to release emergency funds, and to reallocate state agency budgets for emergency response work.

If assistance is beyond their capability, the governor requests a presidential declaration of emergency or major disaster according to the Stafford Act. The governor's request for a major disaster declaration request shall be based on a finding that the disaster is of such severity and magnitude that effective response is beyond the capabilities of the state and the affected local governments and that federal assistance is necessary (Stafford Act Title IV, Section 401). The governor submits an official request to the president through the FEMA regional administrator asking for federal assistance under the Stafford Act. If granted, the governor then appoints a state coordinating officer (SCO) to interact with the federal coordinating officer (FCO) assigned to the disaster operation as the president's representative. A few states, such as Virginia, are also developing State Reservists Programs. The intent behind such programs is "to develop a cadre of trained people who can be called upon during a disaster to support agency response and recovery efforts" (Virginia Department of Emergency Management, 2010).

State National Guard units can play a critical role in responding to a catastrophe. The National Guard has constitutional and statutory roles as both the militia in their respective states and as a federal military reserve force. As the state militia, the governor commands each state's National Guard. Each member of the National Guard has dual status as a member of the National Guard of his or her state and of the army or Air National Guard of the United States, the latter being a reserve component of the army or the air force, respectively.

The National Guard may be activated under state law to deal with civil disturbances or natural emergencies, disasters or catastrophes, maintain vital services (such as hospitals or prisons), conduct drug enforcement operations, and respond to other threats to the security of the state's citizens or violations of state laws. In addition, the president is authorized to activate the National Guard into federal service to deal with a wide range of domestic events. These include suppressing insurrections if requested to do so by the state's governor or legislature; enforcing federal laws are also included.

Governors frequently activate portions of their states' National Guard. Normally a request to use the National Guard at the local level is made by local authorities through the state emergency management agency to the governor or his or her executive agency.

In fiscal year 1993, for example, 34,052 members of the Army National Guard (out of a total of 422,720 members) were called to state active duty in 47 states. They dealt with 148 natural disasters, 40 search and rescue operations, 48 provisions of potable water to communities hit by natural disasters, 13 fires, and 10 law enforcement assistance missions. In recent years, large numbers of National Guard personnel have been called to state active duty to cope with Hurricane Andrew in Florida and Louisiana, and Hurricane Iniki in Hawaii (1992), the Los Angeles riots (1992), and the Midwest floods (1993) (U.S. Senate, 1995, pp. 44–45). Of course, the National Guard also played a substantial role in the response to Hurricane Katrina in the impacted states. We all can remember the television images of the National Guard playing a key role in assisting and protecting the residents of New Orleans following Hurricane Katrina. The Guard remained on patrol in some areas of New Orleans until 2009, a full four years after Katrina struck.

BOX 4.9 STATE GOVERNMENT POTENTIAL USES OF NATIONAL GUARD FORCES

- Search and rescue
- Hazardous materials decontamination
- Civil Support Teams (CSTs) respond to a chemical, biological, radiological, nuclear or explosive (CBRNE) detection
- Communications equipment and personnel
- Transportation equipment and personnel
- Security and maintenance of order
- Mass feeding
- Provision of potable water
- Housing (typically tents)
- Health and medical care
- Sanitation
- Temporary restoration of essential facilities
- Engineering services
- Debris clearance

4.12.3 The Federalist System

As the reader is aware by now, the political system in the U.S. is a shared one among federal, state, and local governments. As a federal system, emergency management involves many stakeholders including our federal executive offices and agencies, the Congress, and state and local elected officials and agencies. Although the federal government provides extensive resources for emergency management and especially support for response and recovery efforts, local governments have the greatest responsibility for emergency management activities and services.

The U.S. federal system dictates as a basic requirement that state and local levels of government must comply with federal legislation, regulations, and court decisions. Thus, state and local jurisdictions must follow the guidelines and frameworks outlined in earlier in this discussion. As a federalist system, emergency management is a decentralized organization that includes many state and local agencies that play a critical role in our response and recovery efforts. As a result of this decentralized system, conflict arises out of the execution of state and local policies and actions.

State, territorial, tribal, and local compliance with federal programs and policies, however, is often due to the incentive of funding requirements that accompany financial resources either for planning/preparedness, response, recovery, or mitigation. Issues associated with control and direction have increased since the formation of the DHS following the attacks of September 11, 2001. The federalist system has advantages including flexibility, diversity, and redundancy of operations, and both local and regional focus on hazards and vulnerabilities (McEntire and Dawson, 2007).

4.13 USE OF NATIONAL GUARD OR OTHER MILITARY FORCES IN CATASTROPHE RESPONSE

The Bush Administration report entitled *The Federal Response to Hurricane Katrina: Lessons Learned* (2006) provided a number of observations concerning the use of the U.S. military in domestic emergency response operations. The military provides a great operational response resource, if utilized properly. Further, the process for obtaining this support is very specific and unfortunately takes some time.

The federal response to Hurricane Katrina demonstrates that the Department of Defense (DoD) has the capability to play a critical role in the nation's response to catastrophic events. During the Katrina response, DoD—both National Guard and active duty forces—demonstrated that, along with the Coast Guard, it was one of the only federal departments that possessed real operational capabilities to translate presidential decisions into prompt, effective action on the ground. In addition to possessing operational personnel in large numbers that have been trained and equipped for their missions, DoD can bring robust communications infrastructure, logistics, and planning capabilities. Since DoD, first and foremost, has its critical overseas mission, the solution to improving the federal response to future catastrophes cannot simply be to "let the Department of Defense do it."

The federal response to Hurricane Katrina highlighted various challenges in the use of military capabilities during domestic incidents. For instance, limitations under federal law and DoD policy caused the active duty military to be dependent on requests for assistance and not pro-active for fear of appearing to violate the Posse Comitatus Act. These limitations resulted in a slowed application of DoD resources during the initial response. Since the president chose not to federalize National Guard units while activating active military units, active duty military and National Guard operations were not well coordinated and served two different commanders, one the president and the other the governor.

While many viewed the deployment of DoD resources in response to Hurricane Katrina as critical to rescue efforts, there remain limitations that need to be acknowledged in the use of military resources in an emergency response. One such limitation is the DoD response authority. For federal domestic disaster relief operations, the DoD currently uses a "pull" system that provides support to civil authorities based on specific requests from local, state, or federal authorities. This process can be slow and bureaucratic. Assigning active duty military forces or capabilities to support disaster relief efforts usually requires a request from a state to FEMA, from FEMA to DoD, an assessment by DoD on whether the request can be supported, approval by the Secretary of Defense or his designated representative, and a mission assignment (a process whereby FEMA agrees to reimburse DoD for its costs) for the military forces or capabilities to provide the requested support. A lengthy process is required from the time a request is initiated until the military force or capability is delivered to the site of the disaster or catastrophe. While this overly bureaucratic approach has been adequate for most disasters to date, in a catastrophic event, the delays inherent in this "pull" system of responding to requests will most likely result in critical life sustaining needs not being met, causing needless pain and suffering. At this point in this book, it should be easy for the reader to imagine a situation in which a catastrophic event in the U.S. would be of such a magnitude that it would require an even greater role for the DoD than in Hurricane Katrina.

The Bush Administration Hurricane Katrina report (2006) also discussed linkages between active duty military units and the state national guard (see below). Worth noting are the potential political struggles that could emerge between the federal government and a governor who controls the National Guard units.

In the overall response to Hurricane Katrina, separate command structures for active duty military and the National Guard hindered their unity of effort. U.S. Northern Command (USNORTHCOM) commanded active duty forces, while each state government commanded its National Guard forces. For the first two days of Katrina response operations, USNORTHCOM did not have situational awareness of what forces the National Guard had on the ground. The military task force assembled to manage the military's response to Katrina, Joint Task Force Katrina

(JTF-Katrina), simply could not operate at full efficiency when it lacked visibility of over half the military forces in the impacted area. Neither the Louisiana National Guard nor JTF-Katrina had a good sense for where each other's forces were located, their capabilities, or what they were doing. For example, the JTF-Katrina Engineering Directorate was not able to coordinate with National Guard forces in the New Orleans area. As a result, some units were not immediately assigned missions matched to on-the-ground requirements. Further, FEMA requested assistance from DoD without knowing what state National Guard forces had already deployed to fill some of the same needs. Also, the commanding general of JTF-Katrina and the adjutant generals (TAGs) of Louisiana and Mississippi had only a coordinating relationship, with no formal command relationship established. This resulted in confusion over roles and responsibilities, delays in delivery of critical services and supplies, and resulted in needless pain and suffering.

4.14 POTENTIAL FEDERAL SYSTEM BREAKDOWNS IN HYPOTHETICAL FUTURE CATASTROPHIC EVENTS

Walters and Kettl (2006) foresees a lack of intergovernmental relationships between federal, state, and local agencies as a potential inhibiting factor to effective planning and response in future catastrophes. They argue that there is a lack of clear roles and responsibilities and decision-making authority, and political infighting will occur. An ever increasing role of the federal government in disaster response and recovery along with an expectation that the federal government will help (actually bail everyone out) continues to erode the sense of responsibility at the state, local, and personal levels. These issues fly in the face of the basic emergency management premise that at least for the first 72 hours, all large-scale disasters are local. States must understand that the key is to strengthen local and state capacity to deal with damaging events to minimize the reliance on federal resources. Some might argue this premise doesn't apply to certain types of catastrophes, such as no-notice, regional megacatastrophes, since the ability of states and local government to mount a response to such a catastrophe will be highly eroded or nonexistent following such a hypercomplex event.

Another potential point of failure is a lack of effective organizational adaptation. Perro's (1984) normal accidents approach and the analysis of high-reliability organizations has provided a good framework in understanding organizations that deal with complex technologies. Perro contends that the production process and how the organization is managed and uses risky technologies make those organizations prone to catastrophic failure. He determined that the structure provided within an organization determines how well adaptation is made in times of crisis. Unfortunately, he saw that too many organizational systems were unable to correct or adapt when needed.

Ward and Wamsley (2007) contend that ICS has been effective as a standardized coordinating system for fighting forest fires and brush fires, but that it is not sufficiently flexible to deal with the diverse disasters or the variety of unexpected issues that typically arise during a catastrophic response operation. While the system created a centralized coordinating structure, it also created an artificial barrier between formal response systems and the informal networks that form at the local level in response to catastrophes. They contend that, in the response to Hurricane Katrina in 2005, the NRP was not fully integrated at the state and local levels negating potential benefits of this approach to organizing for disaster response. Flexibility and adaptation was needed rather that rigid organizational frameworks. Command confusion and lack of coordination were present rather than fluid lines of authority and discretion.

A potential failure point in catastrophic response is when reactive policy making rather than conscious deliberate policy making occurs. Katrina demonstrated that we often remain in a reactive policy-making position and that we then fail to effectively learn from disasters and catastrophes and make constructive adjustments. Kingdon (1984) sees that effective catastrophic policy making is a three-stage process including:

1. *Problem recognition*: This includes the acknowledgement of a situation by policy makers that requires their collective action in the form of new programs, adaptation of existing ones, or the elimination of barriers to a situation. Poor and outdated building codes and the requirement that victims of disasters remain in unsatisfactory housing or public shelters would be two situations requiring policy attention. A sufficient number of policy makers must agree on the nature of the problem for action to be taken to develop a satisfactory solution to address it.
2. *Problem definition*: This includes the agreement by policy makers in a common framework for the issues that have been identified and consensus that action must be taken. Unless a consensus is reached on the nature of the problem, no adequate resolution to it will be made.
3. *Policy adoption*: This is formal lawmaking that can be at the federal, state, or local level. Providing for open debate in federal, state, and local legislative bodies, or full assembly environments provides a basis for adopting measures that will address the problem(s) and result in constituent buy in.

Ward and Wamsley (2007) reflect on the contribution of Baumgartner and Jones (1993) in the theory of "punctuated equilibrium" and its impact on catastrophic planning and response. This theory contends that policy making is an inherently unstable process that is reflected in continued disagreement even following formal approval of a public policy (i.e., critics remain unconvinced of the adopted solution). In this theory, events over time will erode public confidence in the adopted solution allowing the opposition to overturn the previously adopted measures and replace these measures with alternative ones.

Events, such as the response to Hurricane Katrina, open the problem and debate allowing the equilibrium within the policy to become unsettled. This instability provides the opportunity for an alternative policy to be adopted. The U.S. emergency management system is diverse and alternative approaches to dealing with risks are encouraged to account for differences in geography, local culture, or political dynamics. In the unstable environment of emergency management response, in which unprecedented and unanticipated events of national magnitude can exceed the capacity of any given solution, change may be the only constant. This observation is only magnified in catastrophes.

Government decision makers pursue political agendas noting that agencies can and will use rational methods of policy making in an efficient manner. This is in contrast to an approach that distinguishes between political and administrative roles. In an unstable environment, agency heads and their staff may attempt to deal with unstable environments by quickly and efficiently adopting new policies that allow the institution to adapt to existing instabilities within current legislative and policy frameworks, if they exist.

Lack of effective emergency response may lead to political blame games and/or organizational adaptation that overstep current authorities and separation of powers. Unfortunately, the result is still confusion in a very complex environment. The issue remains in how to deal with the confusion in a rational manner that does not lead to conflicts in political and administrative roles in catastrophes.

Ward and Wamsley (2007, p. 219) observe that:

> The myth of an efficient, effective and organized bureaucracy overlooks the complexity, difficulties and near failures involved in achieving our national goals, enabling us to maintain a profound faith in the application of rational governance in pursuit of organizational efficiency. We reject excuses for failure. It is precisely this national consciousness—this myth—that allows the blame game to be so effective. In our rush to find out who or what is responsible for failure, we ignore the complexity of the system and the role that this complexity has played. We substitute simplistic and convenient answers that are grounded on faulty analysis. By identifying the wrong problem or

incomplete one—at best—we formulate a misguided solution. This inevitably results in future and organizational failures.

In a centralized hierarchical system that is reflected in the now defunct NRP, a network approach can acknowledge the contribution that many players at various levels of response provide. The network approach thus provides a means of acknowledging the contribution of different organizational cultures that can work toward a common set of goals and priorities. The goal of intergovernmental and interorganizational cooperation was seriously undermined by response and recovery efforts following Hurricane Katrina. What are needed are high levels of trust, communication, and organizational flexibility in the emergency management system.

4.15 STATE GOVERNMENT LEGAL POWERS DURING A CATASTROPHE

As an example, the document *Powers of the (Louisiana) Governor* (located at: http://www.training.fema.gov/EMIWeb/edu/docs/crr/Session%205%20-%20Appendix%20D%20-%20Powers%20of%20the%20Governor.doc) prescribes what the governor of Louisiana can do in specific situations associated with a disaster or catastrophe. Some of the powers granted to a governor typically include the ability to declare a disaster, order an evacuation, compel an evacuation, limit return of the public to the disaster area, obtain property, suspend state contract bid procedures, or identify routes for an evacuation.

Declaring an emergency activates the emergency response and disaster recovery powers of the state and political subdivisions within a state. Additionally, it typically will activate specific powers for the governor or the designee to take a variety of actions for the duration of the declaration. These may include the following, but are not limited to:

- Deploying and using any forces to which the plan or plans apply.
- Using or distributing any supplies, equipment, materials, and facilities assembled, stockpiled, or arranged to be made available under any law relating to disaster and emergencies.
- Authorizing the governor to act as the commander-in-chief of the organized and unorganized militia and of all other forces available for emergency duty.
- Suspending the provisions of any regulatory statute prescribing the procedures for conduct of government business, or the orders, rules, or regulations of any agency if strict compliance

with any of these provisions would in any way prevent, hinder, or delay necessary action in coping with the emergency.

- Using all available resources of the state or local unit of government reasonably necessary to cope with the emergency event.
- Transferring the direction, personnel, or functions of departments and agencies or units for performing or facilitating emergency services.
- Subjecting to any applicable requirements for compensation, commandeer, or use any private property if the governor finds this action necessary to cope with the emergency.
- Assisting in the evacuation of all or part of the population from any stricken or threatened area in the jurisdiction if the head of the unit of government considers this action necessary for the preservation of life or other disaster mitigation, response, or recovery.
- Prescribing routes, modes of transportation, and destinations in connection with evacuation.
- Controlling ingress to and egress from a disaster area, the movement of persons within the area, and the occupancy of premises in the area.
- Suspending or limiting the sale, dispensing, or transportation of alcoholic beverages, firearms, explosives, and combustibles.
- Making provision for the availability and use of temporary emergency housing.
- At the state level, allowing persons who hold a license to practice medicine, dentistry, pharmacy, nursing, engineering, and similar other professions as may be specified by the governor to practice their respective profession in the state during the period of the state of emergency if the state in which a person's license was issued has a mutual aid compact for emergency management with the state.
- Giving specific authority to allocate drugs, foodstuffs, and other essential materials and services (Indiana Code).

In Tierney, Lindell, and Perry (2001) one of the most challenging aspects of emergency response is the fact that often a decision must be made between options that are universally unattractive. The level of unattractiveness of options is usually elevated in a catastrophe. Sometimes, people impacted by a catastrophe will view the same situation from a very different perspective than public officials. Evacuation is a good example of this phenomenon. Legal authorities for declaring an evacuation are found in state statutes or local ordnances. The law may grant the head of government the authority to force people to evacuate their homes and businesses and may be the best protective step that can be taken. Conversely, the evacuation authority may limit authorities to issuing voluntary evacuation orders. Unfortunately, evacuation can be very expensive and disruptive for households and businesses, especially in coastal areas during tourist season when a hurricane may not follow its predicted path or weaken. Often groups of residents refuse to evacuate for myriad reasons. To be successful, local or state officials may well have to take steps to help overcome the arguments from residents for not complying with the legitimate official orders, such as cost of transportation, housing, food, caring for pets or relatives while evacuated. The key is that state or local officials might have the legal right to impose mandates on citizens, but realize that it is very unpopular and could cause politically negative fallout and resistance from impacted residents.

4.16 STATE LEGAL PROTECTION LAWS FOR VOLUNTEERS

4.16.1 Tort Law

A *tort* is an action that harms another. It occurs when a person acts or fails to act, without right and as a result another is harmed. Torts involve civil actions for personal injuries or property damage rather than a criminal action or a contractual claim. Tort law is defined at the state level by statutes, court decisions, and state constitutional provisions; it applies to government entities, individual citizens, and businesses. The law of torts protects individual and business interests from harm and provides a means for those harmed by another to seek compensation for their loss.

Tort liability claims also provide a basis for distributing losses to those who are responsible for the harm. Tort law thus provides a systematic means for analyzing and resolving liability claims, while protecting both the interests of the person injured and the governmental jurisdiction. Torts encompass a very broad area of the law including:

- *Intentional acts that harm others*: They include trespass, assault and battery, and intentional infliction of emotional distress, defamation, and invasion of privacy.
- *Negligence*: They include unintentional acts or omissions that cause harm to another. Negligence involves an unintentional but wrongful action or inaction by one person, which harms another.
- *Strict liability*: This is liability without fault and relates to situations where one is held responsible for the consequences

of his/her actions or omissions, regardless of fault or exercise of due care. Strict liability was first applied in cases involving abnormally dangerous activities, such as blasting associated with mining and construction, but has achieved significantly broader application in the law of products' liability and workers' compensation (Oleck, 1982).

4.16.2 Negligence

A person has a duty to exercise that degree of care, skill, and diligence that a reasonable or prudent person would exercise under similar circumstances. This rule, as applied to governmental entities, must be understood in terms of the essential elements of negligence.

- *Duty*: The existence of a duty to conform to a defined standard of care either established by statute, defined by common law (based upon judicial decisions), or established by policy by the governmental entity, which is owed to a particular party.
  - A common law duty requires a person to use a reasonable degree of attention, perception, memory, knowledge, intelligence, and judgment in his/her actions.
  - Statutory duties include traffic codes, motor vehicle maintenance codes, workplace safety requirements, construction and maintenance standards, environmental regulations, or inspection requirements.
- *Breach*: A failure to conform to that standard of care, or a failure to carry out the duty.
  - There has been a failure to act according to a standard of care.
  - The courts allow the injured party an opportunity to show that the actions of an individual or business governmental entity or their agent were unreasonable.
- *Damage*: Actual loss or damage to the injured party(ies) (the loss must be real and quantifiable).
- *Causation*: There must be a connection between the act of the individual or business governmental entity or their agent and injury to a third party(ies) and the loss must be related to the act of individual or business governmental entity or their agent.

All negligence cases have these elements in common and absence of proof of any one element will defeat a finding of liability in a court of law.

4.16.3 Immunity from Claims of Negligence for Government Actions

4.16.3.1 Discretionary Immunity

Discretionary immunity evolves from the judgmental decision-making process of public officials and employees. Volunteers that serve in decision-making capacities could enjoy this form of immunity. The key to ensuring that volunteers are protected in catastrophes for their negligent actions is formally appointing the citizen as a representative of the public entity. The volunteer thus becomes a "public actor" rather than a private citizen offering aid. In addition to formally appointing or authorizing the volunteer, the public agency should provide training and guidance to volunteers so as to minimize harm to citizens. A failure to formally appoint the volunteer to assist in catastrophic response thus means that the various forms of "governmental immunity" do not apply to the actions of the volunteer and other standards of care may apply. Note that the various forms of governmental immunity provide extensive protection to public employees, officials, and volunteers even when there is negligence. This protection, however, does not apply where the actor intentionally harms another. Our governmental immunity does not apply to criminal conduct. One further note, the legal considerations would not change depending on the scale of a disaster. Thus, in a catastrophe, volunteers continue to have immunity as in an emergency or a disaster. Based on current legal precedence, the size and nature of the event does not impact the level of immunity.

4.16.3.2 Governmental Immunity

A special form of immunity is recognized in 14 states for some activities of public agencies. State law makes a distinction between governmental functions, which are traditionally performed by the government, and those functions that are proprietary in nature or performed traditionally by the private sector. Under the governmental function theory, core governmental functions, such as public safety, including emergency management, firefighting, police activities, as well as health and building inspections, and the collection of taxes, are mandated responsibilities that can be performed only by governmental entities.

Because of the unique role that these essential governmental functions have in the community, public agencies and employees enjoy immunity from claims of negligence under state law. Each state that recognizes governmental immunity defines what a governmental function is. Proprietary functions, however, have no special immunity attached to the activity. Proprietary activities may be performed by either a public or private organization. Public actors or volunteers do not enjoy immunity when performing proprietary activities.

4.16.3.3 Statutory Immunity

Each state legislature has adopted statutory provisions recognizing immunity in specific public activities even when the actor has been negligent. Immunity from claims of negligence may be recognized for volunteers in catastrophes. These immunity provisions extend protection to negligent acts, but not to actions for gross negligence or intentional actions intended to harm another. Many states have adopted immunity provisions in emergency management activities for anyone acting on behalf of the public authority. A critical element of these provisions involves the defining of an "emergency." If the emergency activity is not included in the definition of "emergency," then the immunity provision does not apply.

4.16.4 Exceptions to Immunity

State statutes may exclude immunity for public officials and duly appointed volunteers in cases of willful misconduct, gross negligence, or intentional harm to others. Willful or wanton misconduct involves highly unreasonable conduct or actions that are an extreme departure from ordinary care. Willful misconduct:

- May involve unreasonable conduct, but is an action that is not taken with the intent to cause injury to another. Gross negligence is more than mere thoughtlessness, inattention, or a mere mistake resulting from inexperience or confusion.
- These actions would be viewed by the courts as negligence and covered by any statutory immunity provisions under state law.

4.16.5 Defenses in Claims of Negligence

Governmental jurisdictions may avoid liability by using two major types of defenses: denial defense and affirmative defense.

- A denial defense directly disputes the allegation by the plaintiff that the defendant has behaved negligently. In effect, the defendant is claiming that he/she (or the governmental entity) has acted with reasonable care.
- An affirmative defense may allow the defendant to avoid liability, even where his/her conduct is negligent. These defenses include, but are not limited to:
  - Statutory immunity (as noted in the above discussion)
  - Settlement of the claim

- Filing or bringing the suit after the statute of limitations (which bars the action)
- Assumption of the risk giving rise to the injury, such as signing a waiver of liability
- The person who is bringing suit contributed to the injury (strict contributory negligence)

4.16.6 Indemnification of Public Employees and Volunteers

Official representatives of a governmental entity, including duly authorized volunteers, who are named individually in a tort action, are generally entitled to protection against personal financial loss or indemnification. This may apply to both attorney fees and judgments that might be awarded against them. Almost all states recognize that the governmental entity is liable for the negligent acts or omissions of its agents, volunteers, or employees who are acting within the scope of their duties as public employees. The employee in this context includes not only paid staff, but also duly authorized volunteers. Elected officials who receive no pay and volunteers would thus be included in this definition of *employee*. The liability for the employee's actions is passed on to the governmental entity as employer, under a theory generally known as "vicarious liability."

The governmental entity may not be liable for an employee's or volunteer's actions if the employee or volunteer acted outside the scope of his/her duties, acted with an intent to harm (malice) or the intent to harm another, or if the actions were with reckless disregard for the rights of others. Most state indemnification statutes provide that, where the employee or volunteer acted with malice or the employee or volunteer's actions were outside the scope of the job, no defense is provided nor judgment paid.

4.17 POLITICAL IMPLICATIONS OF CATASTROPHES AT VARIOUS GOVERNMENTAL AND POLITICAL LEVELS

Catastrophes, such as the challenges like those presented in the response and recovery from Hurricane Katrina, involve government officials, as well as residents, real estate developers, business owners, and many professionals, such as engineers, architects, and urban designers/planners. The debate that results following a catastrophe is one that the entire community has a stake in. Berke and Campanella (2006) pose several questions that are at the core of this debate. Hurricane Katrina provided a "window" on how effective recovery planning can support rebuilding resilient and sustainable communities. Windows are moments of opportunity when a problem has become urgent enough to push for change of entrenched practices (Birkland, 1997). Unfortunately, these windows do not remain open for long and local officials and stakeholders must take advantage of this opening for change. Predisaster planning allows the local community to return by having short-term strategies to cope with temporary housing, damage assessment, debris removal, restoration of utilities, reoccupancy permitting, and reconstruction priorities. Long-term strategies in the predisaster recovery planning include rebuilding moratoria while community resilient and sustainable redevelopment plans are put in place, planning for high-risk areas, and relocation of housing to safer sites, all in an effort to break the event–damage–event–damage cycle that a community experiences. More importantly, precatastrophe recovery planning allows policy makers to envision the community outside of the distress and extreme emotions that are commonplace in the postcatastrophe environment. Given the developments in hazard models, more accurate data, and potential increasing threats brought on by climate change, communities should periodically review their hazards analysis to determine if their conclusions should be altered to address an altered risk from hazards.

Long-term planning must integrate transportation, housing, land use, and environmental issues and balance the social–cultural, economic, and ecological needs of the community. Research has demonstrated that most communities have failed to develop effective predisaster recovery plans. What they have developed has been considered inadequate to address community problems after a catastrophe. The end result is that in the postcatastrophe environment, the community is then engulfed in conflicts over recovery policies and expenditure of limited resources (Burby and May, 1997; Mader, 1997; Nelson and French, 2002).

Policy makers at the national, state, and local level have the opportunity to address long-term issues that involve community and infrastructure resiliency and vulnerability to hazards. During the response and recovery period to a catastrophe, policy makers may well be more concerned with dealing with immediate concerns, such as the restoration of public services, than with long-term issues that will impact the community for an extended period, such as deciding whether to allow the reconstruction of homes and businesses in high risk areas including floodplains.

Smith (1996) also observes that disasters and, by extension, catastrophes, provide opportunities for political change. He sees that catastrophic disasters present a chance for new alliances to emerge and mobilize. A catastrophe will almost certainly shape, maintain, destabilize, or destroy both political organizations and relations. They create contexts in which power arrangements may be articulated and challenged that change political perceptions, shape individual intentions, and strengthen or dissolve institutional alliances. A future catastrophe in the United States will certainly bring about new ideologies, activism, and power alliances.

Berke and Campanella (2006) suggest three key issues that must be addressed in shaping the rebuilding of a resilient community that has suffered a catastrophe:

1. Some local, state, and federal land use and development policies foster rebuilding or avoiding development in hazardous areas and impede sensible local predisaster planning. Gulf coast states, with the exception of Florida, fail to require comprehensive land use planning in coastal areas. As a result, communities often rebuild without consideration of future risks from well-known hazards, such as riverine floodplains, hurricane coastal high hazards areas, and known seismic fault areas. In the wake of Katrina, most states along the Gulf Coast have failed to support or require comprehensive local planning to direct rebuilding away from high-risk areas. The federal government, including Congress, has also failed to require local planning to avoid development in high hazard areas. It is worth noting that in 2009 a senator from a Gulf Coast state held up the confirmation of the FEMA Administrator to pressure FEMA not to issue a new Flood Insurance Rate Map (FIRM) that would have made it more difficult for people to rebuild their Hurricane Katrina-destroyed homes in a Coastal High Hazard Area flood zone.
2. New urban models must be utilized to enhance resiliency and avoid risk from high hazards. Urban sprawl, rather than high-density new urbanism, normally reflects development following a catastrophe. Compact urban form design must be considered to relocate structures out of harm's way as communities struggle to recover from catastrophes, yet house the prevent population in safe manner.
3. Engaging community stakeholders in the planning process in a meaningful manner is very important. Citizen participation must be viewed as a means of restoring and repairing the social fabric of a community following a catastrophe. Engaging the entire community in a meaningful manner provides opportunities for the community to heal in many ways. Research suggests that when the community is not engaged, plans and implementation strategies do not benefit from local knowledge and capacities (Healy, 1997; Zaferatos, 1998). All too often, external experts who lack local understanding of conditions dominate policies. The end result is that the planning that takes place actually creates opposition to and conflict with to the plans.

> Catastrophes present an additional dilemma for local officials who want to engage displaced residents in the long-term recovery process. The degree of displacement following Hurricane Katrina illustrates the challenges that local, state, and federal officials face in a catastrophe. A strategy that works in a local or regional disaster might work in a large-scale catastrophe where residents have be dislocated from their homes and have been disbursed into temporary housing located across a large area, such as what resulted for residents of the greater New Orleans area in Hurricane Katrina.

Whether the above strategies will work in a catastrophe or not, there are certainly lessons to be learned that can apply to a catastrophe. Some argue that there is too much at stake for the federal government to stand by and wait for state and local governments to embrace strong land-use requirements in mitigating high hazards. Some suggest that the federal government must take strong measures, i.e., use the *power of the purse*, to require the following if states, territories, tribes, and local governments want to gain access to federal funding to support the recovery, reconstruction, and mitigation phase of a catastrophe:

- Require state and local governments to adopt classic grass roots organizing to encourage citizen participation and the renewal of civic institutions that can participate in difficult recovery problems.
- Provide opportunities for communities to acquire sound civic planning skills and ensure that all parts of the community are engaged.
- Ensure that all parts of the community are engaged.
- Reinforce the value of collective action rather than a culture of quick action without the consideration of the broad networks resident within the community.

For further discussion of local government organizational response issues and strategies, see Tierney et al. (2001).

4.18 LEGAL ISSUES INVOLVING GOVERNMENTAL POWERS FOR CATASTROPHE RESPONSE AND RECONSTRUCTION

4.18.1 Use of Private Resources in a Disaster Response

State emergency management laws usually provide for acquiring private property in a declared disaster, which would include a catastrophe. It should be noted that this provision affords a constitutional protection concerning takings that require the public entity to provide "just" compensation to the property owner for the use of or taking of their property during response to an emergency. Public agencies should have a standard operating procedure for taking private property in a catastrophic response, including consultation with agency legal counsel and notice to property owners when possible. In cases where time does not permit or other property is in danger, documentation of the existing situation is advisable to provide for a strong justification for the public agency actions. Many disasters provide clear illustrations of when property can be taken, at least temporarily, to aid in the response or in order to remove barriers to an effective response.

4.18.2 Communicating Warnings

Many public officials, as apolitical calculation, are hesitant to issue clear warnings when they believe that the public is in danger out of fear of making the wrong decision. The law in each state provides "discretionary" immunity to public policy makers who are charged with executing the laws and protecting the property and lives of the public. Where public officials must determine public policy, this form of governmental immunity protects them. Issuing a warning would be an example of a public policy decision that falls under the protection of governmental discretionary immunity.

Many public officials rely on the expertise and experience of professional staff in providing background information for the determination of a public policy decision. Where state or local officials are specifically allowed to issue public hazard warnings and evacuation orders and they rely on the information and recommendations of their staff, the discretionary immunity still applies. For the public employee, they are also acting in their official capacity and enjoy protection from civil liability suits. In this situation, the employer, i.e., the government agency, is responsible for the actions of the employee (vicarious liability) and will defend the employee along with the agency if a claim is filed in court.

The fact that employees form their recommendation from computer hazard models, experience, or direct observation of a hazardous scene does not impact the protections that they enjoy under state law. It is worth noting that there are many factors that influence public policy decisions involving warnings. Economic, political, natural, and social factors all affect the determination to order an evacuation. The fact that the public official made a poor decision and failed to issue a timely warning does not remove the protections provided to him/her under state law. The impacts from the poor decision may conclude in political changes and actions well after the catastrophe that do not involve the courts in civil actions.

4.18.3 Prohibiting Access to Damaged Areas

State emergency management laws and regulations generally allow a governor to limit or exclude the public from entering a damaged area. Protecting the public in an unsafe environment also may be viewed as being under local government police powers. The inconvenience on members of the public to gain access to disaster areas is well justified by such dangerous circumstances. Even if citizens make a constitutional claim against local or state officials, the courts would likely see that protecting the public in an unsafe circumstance was a small price to pay in order to allow emergency responders to perform their job without the distraction provided by the public who just want to know what is going on. As with other situations, it is advisable for public officials and senior managers to document the situation that they face, such as by taking photos of the area to show that the area is unsafe and that protective actions are in the public interest.

4.18.4 Compelling an Evacuation

Requiring the public to leave a high-risk area when a catastrophe is likely to occur is not always provided under state law. Public officials may have even less authority to force an evacuation from an area after a catastrophic event has rendered the area unsafe for continued human occupancy. Public officials must work closely with their legal counsel to confirm that they have the authority to remove citizens from a high-risk area both pre- and postevent. Knowing precisely the applicable legal authority is a wise approach and legal counsel will appreciate the opportunity to clarify any provisions in the law that could influence when and how an evacuation might be initiated.

4.19 POLITICAL AND LEGAL CHANGE THAT CAN RESULT FROM CATASTROPHES

Some believe that Hurricane Katrina revealed many flaws in catastrophic planning that included:

- Failing to reduce future vulnerability of structures to hurricane winds.
- Failing to examine the inefficient use of local resources in the courts and administrative offices.

- Failing to link priorities for recovery efforts with land use decisions.
- Failing to provide additional financial support above flood insurance for residential structures rather than providing financial assistance to restoration of rental units (which provide rental opportunities for displaced families).
- Failing to explore liability claims for extensive property damage perceived to be the fault of local, state, and federal agencies.

The governor of Louisiana called the state legislature into special session following Hurricane Katrina to establish a strong statewide building code (see Box 4.10). Political, economic, legal, and social forces led to the adoption and implementation of a comprehensive and rigorous statewide building code. The catastrophe revealed the need for strong building codes to support the recovery process to break the event–damage–event–damage cycle. Local and state officials viewed the consequence of failing to adopt a strong building code and determined they were unacceptable. Mortgage lenders, property casualty insurers, and business interests drove the push for changes in the law because they believed that a sound recovery had to be based on safe construction guidelines from the new building code that would result in more resilient communities.

BOX 4.10 IN THE WAKE OF HURRICANE KATRINA, LOUISIANA ADOPTS NEW BUILDING CODE

December 2005—Louisiana Governor Kathleen Blanco signed a bill last week that calls for the state to adopt the International Building Code (IBC), International Existing Buildings Code (IEBC), International Residential Code (IRC), International Mechanical Code (IMC), and the International Fuel Gas Code (IFGC).

The bill applies to buildings rebuilt in the wake of Hurricanes Katrina and Rita, and to all buildings built or rebuilt statewide starting in 2007. Under the legislation, the 11 parishes hit hardest by the hurricanes must put the new code into effect in 30 days if those parishes already have inspectors. If they do not, they have 90 days to begin enforcement. The bill also establishes a 19-member council to oversee enforcement of the codes by local governments.

While Hurricane Katrina devastated much of Louisiana, the state is poised to rebuild stronger and safer than ever using the International Codes (I-Codes) developed by the International Code Council.

In a special session, the state legislature approved adoption of the International Building, Residential, Existing Building, Mechanical

(continued on next page)

and Fuel Gas Codes for use in Louisiana. The bill applies to buildings rebuilt in the wake of Hurricanes Katrina and Rita. It also will be required for all buildings built or rebuilt statewide starting in 2007. Under the legislation, the 11 parishes hit hardest by the hurricanes have up to 90 days to begin implementing and enforcing the wind and flood provisions of the International Building and Residential Codes. The code requires homes and businesses built along the Gulf Coast to withstand winds of 130 to 150 miles per hour. The bill also establishes a 19-member council to oversee enforcement of the codes by local governments.

"The massive effort to rebuild Louisiana will be long and difficult. However, with the International Codes in place to help guide reconstruction, homes and businesses will be safer, stronger and more resistant to future natural disasters," said Sara Yerkes, International Code Council Senior Vice President of Government Relations. "As we have witnessed, in addition to the loss of life, there are many repercussions when natural disasters damage homes and businesses. Hurricane damage disrupts private industry and government services, puts people out of work, reduces disposable income and diminishes the tax base. By adopting and enforcing I-Codes, the state is helping to protect lives and property while limiting the far-reaching effects of hurricanes and other natural disasters."

Many states, including hurricane-prone states, enforce the I-Codes or state codes based on the I-Codes (such as the Florida Building Code), for residential and commercial buildings. I-Codes contain the latest technologies for building construction. They take into account valuable lessons learned over the years. I-Codes provide state-of-the-art requirements for hurricane resistance, based on wind speed data collected from previous hurricanes. In wind-borne debris regions, I-Codes address window, garage and door protection, such as shutters and impact-resistant windows, to protect against flying debris. I-Codes also provide wind load criteria for the design of hurricane-resistant roof tie-downs and exterior cladding.

"Though there may be a slightly higher initial cost, homes and commercial buildings constructed under the I-Codes are less likely to be destroyed during a natural disaster, greatly reducing costs to the property owner. The added level of protection for your home, your belongings, and, most importantly, your family will pay off in the long run," said Yerkes. "Properly constructed buildings and homes are more resistant to general deterioration as well." From: http://www.bookmarki.com/Blanco_Signs_Louisiana_Building_Code_Bill_s/190.htm.

During the initial post-Katrina recovery process, efforts were initiated by the mayor's office to ensure that the City of New Orleans had a sound and stable legal and administrative system. Within months of the catastrophe, changes were made in the state constitution to address key local problems and merge small tax assessor offices into a single parish unit. Information concerning individual property was needed from the assessor's offices for the recovery process and was unfortunately located in offices throughout the city. In addition, initiatives to consolidate law enforcement offices were also undertaken following Katrina. The changes would probably have been unthinkable before Hurricane Katrina as each of these offices represented political power and many political patronage jobs.

Recommendations to reduce the vulnerability of individual properties to flooding were provided in the initial assessment as part of establishing a rebuilding plan for the City of New Orleans. The mayor's office released the Urban Land Institute's comprehensive recommendations in "Bringing Back New Orleans" for review and discussion in late 2005 (see: http://www.uli.org/ResearchAndPublications/Reports/Advisory%20Service%20Panel%20Reports/~/media/Documents/ResearchAndPublications/Reports/AdvisoryServicePanelReports/NOLA_Conculsions.ashx). Many residents saw that their neighborhoods would not be included in the recovery plan as a priority. Their negative comments pushed the mayor to reject the Institute's recommendations. A multiyear planning process was then initiated that has allowed residents to rebuild where they wish. At the time of writing, no real priorities appear to have been established. A comparison of the initial Urban Land Institute Plan (2006) with the City of New Orleans' current recovery plan provides a basis for examining how the political environment has dominated recovery in the city to the detriment of sound rebuilding policy in the city's highest risk neighborhoods, which is where most the city's at-risk population resides.

The Louisiana Recovery Authority, the governor of Louisiana and the Louisiana Congressional delegation proposed using funds provided by the U.S. Congress to establish the Road Home Program. The funds enabled over 150,000 homeowners to obtain financial support to rebuild, relocate in the city, within the state, or secure a buyout of their residence. Limited funds were provided for restoring or rebuilding rental property. This program was generally perceived by outside observers and the media as being rife with bureaucratic delays, too much required paperwork, and only being marginally successful. Examine the Road Home Program further at http://lra.louisiana.gov/.

The key to building or rebuilding and sustaining resilient communities is the establishment of sound hazard mitigation policies that are reflected in local ordinances and development plans and required by local, state, and federal laws, regulations, and policies. Strong building codes and land use planning laws provide the legal framework for ensuring that our communities are rebuilt to be resilient following a catastrophe. Priorities that include expenditures in retrofitting at-risk critical infrastructure reflect a political commitment to reducing the adverse economic and social impacts from future catastrophes.

Previously we noted that there is a window of opportunity for political change present in a catastrophe. Pelling and Dill (2006) note this socio-political dynamic, and the watershed moment presented by a catastrophe that provides for social political action at local, state, and national levels. They observe that policy decisions involve social equity, justice, vulnerability, power relations, and environmental change. Again, political change may include new alliances and leadership.

Catastrophes do not cause political change, rather they act as catalysts that put into motion social processes at different social levels. The political change is thus the result of the precatastrophe socio-political and cultural characteristics of the local, state, or national scene. Catastrophes occur in a social ecological context where some social organizations flourish and others fail and where specific types of relationships with external power affect local, state, and national conditions. Catastrophes will probably occur across multiple political jurisdictions, catalyzing regional political tension. Catastrophes are often a product of development policies that will be open to scrutiny by the media, as well as dominant political and institutional systems. Existing inequalities can be exacerbated by postcatastrophe governmental manipulation. The kind of political relationships that exist between sectors before the catastrophe will largely predicate the way in which the state and other sectors will act in a catastrophe planning and response. Political regimes have been known to interpret spontaneous collective actions by nongovernment sectors in the aftermath of a catastrophe as a threat and respond with repression. In the aftermath of a catastrophe, political leaders may regain or even enhance their popular legitimacy. The repositioning of political actors in the aftermath of a catastrophe unfolds at various levels of government.

Pelling and Dill (2006) conclude that political outcomes are largely influenced by existing social conditions and contracts. Suppressed values and arrangement can reemerge, or become further entrenched as a result of a catastrophe. During the recovery, power alignments can lead to the manipulation of external resources and distribution of economic power. Where new forms of organization become too effective, they may be perceived as a challenge by existing institutions. It is here that democratic and authoritarian regimes tend to differ in their strategies for survival.

Opportunities for change following a catastrophe may provide the political climate to address issues and complex problems that under normal circumstances could not be addressed. Pratt (2006) raises the question of whether the catastrophe resulting from Hurricane Katrina was the exception or if it is just an example of our failure to bring about constructive changes following a catastrophe. Was Katrina the exception? Was it unforeseen and did it exceed our comprehension? He notes that Hurricane Betsy ravaged the New Orleans area in 1965 and over the next 40 years the city's growth expanded into low-lying very vulnerable parts of the city. Pratt suggests that we view this as an "event" that provides a framework for examining our social, economic, political, and environmental systems.

Vale and Campanella (2005) raise a similar issue in noting that most modern catastrophes that impact a large city result in rebuilding at the same location and have yet to see a city entirely relocate. One may view Katrina's impact on New Orleans not from a single event, but a cluster of traumatic episodes that include socioeconomic decay, diminishing investment in infrastructure and buildings, large-scale abandonment, population flight, and decay reflected in schools, water and sewer systems, transportation, and healthcare over an extended period. They also point out that political and legal systems are not isolated from the social, cultural, and economic elements of the city and, thus, are part of them.

Birkland (2006) argues that "focusing events," such as Katrina, elevate problems on the policy agenda, thus gaining "mass and elite attention." Many of the policies may have been ignored or not expressed prior to the event, but afterward they become more acceptable from a political perspective. New Orleans Mayor Nagin survived reelection in the spring of 2006 and was successfully reelected mayor. Governor Blanco chose not to seek reelection after completing her term. Political polls gave her little chance of successfully running for reelection. Birkland notes that disaster events, and by extension catastrophes, may not lead to expected political outcomes or policy changes because the event may overwhelm everyone so as to limit opportunities to learn. The attention that is generated and the learning that results is a function of the consequences of the event.

Political debate is generated by significant events so that policy change or legislation is triggered by these events. But Birkland sees that new ideas are not necessarily developed in response to significant events, but that these events reinvigorate attention to preexisting ideas and propositions. Finally he argues that disasters and catastrophes are not new, but that catastrophic events help move national policy. The "learning" occurs over time from many events and the large-scale event, such as a catastrophe, thrusts the new policy into place.

Media coverage of a catastrophe will drive major policy considerations and learning as well. The presence of the media promotes the consideration of a policy change that may be articulated in the form of new legislation (Birkland, 2006). The presence of a major policy coalition contributes to the likelihood that the policy will be adapted. Focusing events, such as catastrophes, tend to break up logjams that result from political stalemates. Positions that in the past are held tightly suddenly become more adaptable because the event demands action.

4.20 CHALLENGES OF INTERJURISDICTIONAL PARTNERSHIPS

4.20.1 Illusion of Partnerships

Katrina appears to have dispelled any illusion that the answer to interorganization collaboration was solely systematic partnerships. The collaboration unfortunately needs more than just an interagency agreement. One could argue that Katrina was an indictment of existing arrangements that were supposed to provide a basis for interagency coordination and cooperation. One must acknowledge that we, in the United States, have not experienced, in modern times, a catastrophe of the scope, scale, and combination of effects that followed in the wake of Katrina. The question is what went wrong with our interjurisdictional collaborative arrangements? Some just classify Katrina as a "megacatastrophe" (King, 2005; Sylves, 2005; Litan, 2005), but based on most measures of a catastrophe, Katrina was basically a lower intensity catastrophe if it was one at all.

Mitchell (2006) believes that we fail to see the value in taking a "holistic" approach in dealing with disasters and by extension catastrophes. As an example, federal, state, and local plans were not fully integrated, leading to many of the failures observed in Hurricane Katrina. Too often, we approach catastrophe management from a local scale not acknowledging our interdependency between local, regional, state, national, and international levels that is accentuated in a catastrophe. Unfortunately, our collective response to Katrina may be characterized as simply "muddling through," without any significant real change.

Platt (2000) notes that the U.S. Congress has created a complex array of federal agencies based on laws, agency programs, funding, policies, and strategies that are intended to operate on the basis of partnerships, with state and local agencies, nongovernmental organizations, and the private sector. Unfortunately Congress has passed over 50 different laws and President George W. Bush issued numerous executive orders to authorize this patchwork emergency management system. This complexity is antithetical to flexible operation that is needed in an effective response to a catastrophe.

4.20.2 Role of Interorganizational Partnerships

Partnerships are a policy instrument for building collaborations between entities and they provide us a more holistic strategy to deal with cross-political boundary, large-scale event response, mitigation, and recovery issues. Partnerships refer to mutual collaboration and shared responsibility among groups that have common goals and they require broad-based interjurisdictional thinking and decision making. Given that much of the public safety governmental role in the United States is locally driven, partnerships are at the heart of the U.S. emergency management policies.

4.20.3 Interjurisdictional Partnerships Issues

Most interjurisdictional partnerships are interest-centered and tend to last only as long as the groups that come together share those interests (Mitchell, 2006). Once those shared interests change, the partnerships weaken or disappear. Because catastrophes tend to occur infrequently, partnerships that focus on catastrophic issues may suffer this fate. Partnerships then must be based on something more stable than just mutual interests. Vertical and horizontal partnerships are based on different incentives. For our federal–state alliances, these vertical linkages are centered on federal support and compensation that is beyond the capacity of state and local entities. Funding drives cooperation and collaboration. Horizontal cooperation has a very different basis and is usually goal-directed or need-based.

Jurisdictions that adopt mutual aid agreements should be aware of potential legal claims that may arise from doing so. While agreements address liability for damage to third parties or for their employees, state law must be examined to ensure that each party in the agreement is protected. Most state disaster acts provide immunity for agencies with formal partnerships that are recognized under state law (Mitchell, 2006). More complex collaboration between agencies may be required in a catastrophe and may well be beyond the scope of current statutory protections.

4.20.4 Strategies for Successful Interjurisdictional Partnerships

Mutual aid agreements are a key part of formalizing partnerships between jurisdictions and agencies. Their content is guided by a number of standards including National Fire Protection Association (NFPA) 1600 and NIMS. The elements of these agreements may be found on the Emergency Management Assistance Compact (EMAC) Web site (www.emacweb.org).

BOX 4.11 ELEMENTS OF AN EMERGENCY MANAGEMENT ASSISTANCE COMPACT

- Definitions of key terms used in the agreement
- Roles and responsibilities of individual parties
- Procedures for requesting and providing assistance
- Procedures, authorities, and rules for payment, reimbursement, and allocation of costs
- Notification procedures
- Protocols for interoperable communications
- Relationships with other agreements among jurisdictions
- Workers compensation
- Treatment of liability and immunity
- Recognition of qualifications and certifications, and sharing agreements, as required

Despite some of their membership who believed that there was no need to establish a partnership, Mitchell (2006) notes that the European Union directed and implemented a continent-wide integrated hazards response strategy. A critical part of this strategy is a comprehensive education and communication effort to acquaint agencies and nations with the benefits of such an interinstitutional arrangement. The European Union approach illustrates a more holistic one in contrast to the U.S. effort that is more decentralized. Adoption of broad-based goals, such as sustainable development in which multiple jurisdictions share, is critical. Sustainability combines economic, social, and ecological outcomes and forms the basis for a shared vision of the future for the region.

4.21 DISCUSSION QUESTIONS

1. Who has primary control to declare a catastrophe?
2. What is the role of the federal government in catastrophes?
3. What is the primary benefit for obtaining a federal declaration of a disaster?
4. Hurricane Katrina provided at least a partial test of federalism and the capacity of the U.S. emergency management system to deal with a catastrophe. What do you see that Katrina showed us about federalism and catastrophes?

5. The Posse Comitatus Act (18 U.S.C. 1385), along with other related laws and administrative provisions, prohibits the use of the military to execute civilian laws unless expressly authorized by the Constitution or an act of Congress. What is the intent behind the limitations imposed on the federal government? What do you see are the impacts that this imposes in a catastrophe?
6. What potential system breakdowns could limit effective preparedness and response efforts to catastrophic events? How should these system limitations be addressed? Are these limitations also seen in cross boundary international catastrophic events?
7. The White House, Senate, and House of Representatives all examined federal system breakdowns with Hurricane Katrina and Rita. Review the three assessments of the Hurricane Katrina and Rita responses and identify common understandings of the nature and workings of the system at the time of the disasters.
   - Do they agree on common problems or situations?
   - Review each assessment and compare the federal system responses to one another. How did the executive agency assessment compare to the two legislative assessments?
   - How do you account for the differences in the three reports and their views of system agency performance?
8. What other powers does a governor typically have in a disaster?
   - Are the powers limited and if so how?
9. What is the best approach in defending against a claim of negligence for volunteers?
10. What does citizen engagement in recovery planning contribute to full restoration of the community?
11. How does the local community engage the displaced population in a meaningful planning process?
12. What policy issues are present in encouraging displaced people to engage in the recovery planning process and in encouraging them or discouraging them in returning to their former community?
13. In a catastrophe, how should policy makers address whether the local population should be encouraged to return and under what conditions?
14. How did political forces at all levels of government influence implementation of emergency response plans in Katrina?
15. What makes catastrophes so different that our normal legal framework for disasters will not fit the situation?

FURTHER READINGS

General Accounting Office. 1993. Disaster management: Improving the nation's response to catastrophic disasters (GAO Report RCED-93-186), July 23. Washington DC: GAO. Online at: http://archive.gao.gov/t2pbat5/149631.pdf (accessed January 10, 2010).

Government Accountability Office. 2006. Emergency preparedness and response—Some issues and challenges associated with major emergency incidents: Statement of William O. Jenkins, Jr., director, Homeland Security and Justice Issues.

Homeland Security National Preparedness Task Force. 2006. Civil defense and homeland security: A short history of national preparedness efforts. Washington, D.C.: DHS.

Jenkins, W.O. Jr. 2007. *Homeland security: Preparing for and responding to disasters*. Washington, D.C.: Governmental Accountability Office (GAO-06-467T). Online at: http://www.gao.gov/new.items/d06467t.pdf (accessed January 10, 2010).

Quarantelli, E.L. 1994. Catastrophes are different from disasters: Some implications for crisis planning and managing drawn from Katrina. Newark, DE: Disaster Research Center (DRC), University of Delaware.

Tobin, G.A. and B.E. Montz. 1997. *Natural hazards: Explanation and integration*. New York and London: Guilford Press.

U.S. Congress. House. *Volunteer Protection Act*. 1997. 42 USCA. §§ 14501–14505. 105th Cong., June 18.

REFERENCES

Baumgartner, F. and B. Jones. 1993. *Agendas and instability in American politics*. Chicago: University of Chicago Press.

Berke, P.R. and T.J. Campanella. 2006. Planning for post-disaster resiliency. *The Annals of the American Academy of Political and Social Science* (AAPSS) 604: 192.

Birkland, T.A. 1997. *After disaster: Agenda setting, public policy, and focusing events*. Washington, D.C.: Georgetown University Press.

Birkland, T.A. 2006. *Lessons of disaster: policy change after catastrophic events*. Washington, D.C.: Georgetown University Press.

Burby, R. and P. May et al. 1997. *Making governments plan: State experiments in managing land use*. Baltimore: Johns Hopkins University Press.

Bush Administration. 2006. The federal response to Hurricane Katrina: Lessons learned. Washington, D.C.: GPO.

Davis, L.E., J. Rough, G. Cecchine, A. Gereben Schaefer, and L.L. Zeman. 2007. *Hurricane Katrina: Lessons from army planning and operations*. Santa Monica, CA: Rand Corporation.

Department of Homeland Security. 2008. *The National Response Framework*. Washington, D.C.: DHS.

Dror, K. 1988. Visionary political leadership: On improving a risky requisite. *International Political Science Review* 9(1): 7–22.

Farber, D. A. and J. Chen. 2006. *Disasters and the law*. New York: Aspen Publishers.

Healy, P.. 1997. *Collaborative planning: Shaping paces in fragmented societies*. London: Macmillan.

King, R.O. 2005. Hurricane Katrina: Insuring losses and national capacities for financing disaster risk. September 15. CRS Report for Congress, no. RL 33086. Washington, D.C.: Congressional Research Service, Library of Congress.

Kingdon, J. 1984. *Agendas, alternatives and public policy politics*. New York: Harper Collins.

Litan, R.E. 2005. Sharing and reducing the financial risks of future megacatastrophes. Economic Studies Working Paper, November 11. Washington, D.C.: Brookings Institution.

Mader, G. 1997. Enduring land use planning lessons from the 1971 San Fernando earthquake. *Earthquake Spectra* 13(4): 721–733.

McEntire, D.A. and G. Dawson. 2007. The intergovernmental context. In *Emergency management: Principles and practice for local government*, 2nd ed., eds. W.L. Waugh, Jr. and K. Tierney, 57–70. Washington, D.C.: ICMA Press.

Mitchell, J.K. 2006. The primacy of partnership: Scoping a new national disaster recovery policy. *The ANNALS of the American Academy of Political and Social Science* 604: 228.

Nelson, A. and S. French. 2002. Plan quality and mitigating damage from natural disasters. A case study of the Northridge earthquake with planning policy considerations. *Journal of the American Planning Association* 68:194–207.

Oleck, H. 1982. *Oleck's tort law practice manual*. Englewood Cliffs, NJ: Prentice-Hall.

Pelling, M. and K. Dill. 2006. *Natural disasters as catalysis of political action*. New York: Chatham House. ISP/NSC Briefing Paper 06/01, 4–6.

Perro, C. 1984. *Normal accidents: Living with high risk technologies*. New York: Basic Books.

Platt, R. 2000. Extreme natural events: Some issues for public policy. Paper presented at Extreme Events Workshop, Boulder, CO, June 7–9.

Pratt, L. 2006. New Orleans and its storm: Exception, example, or event? *American Literature History* 19:251–265.

Smith, K. 1996. *Environmental hazards: Assessing risk and reducing disaster.* 2nd ed. London and New York: Routledge.

Sylves, R.T. 2005. Revolution needed in U.S. emergency management, EIIP (Emergency Information Infrastructure Project) Virtual Forum Presentation. September 15. Online at: http://www.emforum.org/vforum/lc050914.htm (accessed Januar 10, 2010).

Tierney, K.J., M.K. Lindell, and R.W. Perry. 2001. *Facing the unexpected: Disaster preparedness and response in the United States.* Washington, D.C.: Joseph Henry Press.

Urban Land Institute. 2006. *A strategy for rebuilding New Orleans.* Washington, D.C.: Urban Land Institute.

U.S. Code (2002). Chap. 1. Homeland Security Organization Section 466. Sense of Congress reaffirming the continued importance and applicability of the Posse Comitatus Act.

U.S. Congress. House. Select Bipartisan Committee to Investigate the Preparation for the Response to Hurricane Katrina. 2006. *A failure of initiative.* 109th Cong., 2d S.

U.S. Senate Bipartisan Task Force on Funding Disaster Relief. 1995. Federal Disaster Assistance: Report of the Senate Task Force on Funding Disaster Relief. 104th Congress, first session, Doc. 104-4.

Vale, L. and T. Campanella, eds. 2005. *The resilient city: How modern cities recover from disaster.* Oxford, U.K.: Oxford University Press.

Virginia Department of Emergency Management. 2010. VDEM Reservists. Retrieved from: http://www.vdem.state.va.us/programs/reservist/index.cfm.

Walters, J. and D. Kettl. 2006. The Katrina breakdown. In *On risk and disaster: Lessons from Hurricane Katrina.* eds. R.J. Daniels, D.F. Kettl, and H. Kunreuther. Philadelphia: The University of Pennsylvania Press.

Ward, R. and G. Wamsley. 2007. From a painful past to an uncertain future. In *Emergency management: The American experience 1900–2005.* Ed. C.B. Rubin. Fairfax, VA: Public Entity Risk Institute.

Zaferatos, N. 1998. Planning the Native American tribal community: Understanding the basis of power controlling the reservation territory. *Journal of the American Planning Association* 64(4): 395–410.

II Conclusion

Section II introduced you to the inherent ethical, political, and legal issues of catastrophic events. With respect to ethics, this section addressed the main ethical and value dilemmas and quandaries that will likely be faced before, during, and after a catastrophe. We also explored the legal framework associated with government response to catastrophes that includes the use of the military for domestic response, suspension of civil rights, and federal control of industrial output as well as the states' role in delivering an effective response. Political factors as well as organizational dynamics were included to provide a basis for understanding the complex environment in which preparing for extreme events may take place. We learned that literature in the fields of political science and public administration provides us with insights into conflicts that arise in highly stressful events and the nature of the problems that evolve from our attempts to deal with a catastrophe.

With respect to ethics in catastrophes, an introduction to catastrophic ethics was offered, followed by various definitions of ethics that shed some insight into this highly complex matter. Ethical duties related to professional roles were explored, including whether emergency management is a profession and, if so, do emergency managers have a higher order of ethical duties. This discussion included a presentation of the International Association of Emergency Manager's *Code of Ethics*.

We learned about the moral community and how is it defined in terms of catastrophic response and readiness. We then assessed the competing ethical theories and frameworks. These frameworks included utilitarian catastrophic response, deontological perspectives, duties and principles to govern catastrophic planning and response, environmental ethics, and virtue ethics, all as they relate to catastrophic response. Our exploration of ethics ended with a discussion of ethical dilemmas that nongovernmental organizations face in responding to and assisting with the recovery and reconstruction following a catastrophe.

Section II also discussed, in considerable depth, the political and legal issues associated with catastrophes. With respect to the U.S. emergency management system, the FEMA Comprehensive Preparedness Guide, National Incident Management System (NIMS), and National Response Framework (NRF) were all discussed as they relate to a catastrophe.

The intra- and intergovernmental emergency response framework was assessed in an attempt to explore whether the existing U.S. emergency management system can be effective when faced with a catastrophe. We learned about the principles of U.S. federalism and exceptions to federalism were explored. The value of intergovernmental collaboration was assessed and we found that such collaboration is the cornerstone of effective catastrophic planning and response.

Federal government plans for catastrophe readiness and response ensuring enduring federal and state constitutional governments were presented. We learned that these "doomsday plans" have been around since the early stages of the cold war and have recently modernized following the events of 9/11. The role of the military was discussed, including the use of the National Guard or other military forces in catastrophe response. In this discussion, the limits of *posse comitatus* were discussed as well as the potential federal system breakdowns in hypothetical future catastrophic events. State government legal powers during catastrophes were presented, along with state legal protection laws for volunteers. We concluded Section II by looking at the political implications of catastrophes at various governmental and political levels and we explored the political and legal change that can result from catastrophes.

SECTION III

Operational Issues

- Overview
  Section III explores the postcatastrophic operational environment that an emergency manager might experience. This part is not meant to be an exhaustive discussion of postcatastrophic operational issues, but instead focuses on how these issues differ during a response to a catastrophe. Many, if not all, environs will be altered by a catastrophe. Government, industrial, health and safety processes, critical infrastructure, and other systems will cease to operate as intended for extended periods of time. This part of the book will describe the impact a catastrophe might have on logistics, critical infrastructure, mass care, as well as mass evacuation and relocation.
- Learning Objectives. By the end of this part, the reader should be able to:
  - Compare current potential failure points in the national logistics system and resource mobilization that might result from a catastrophe.
  - Correlate the roles and limits of technology in catastrophe response.
  - Correlate the nexus between logistic systems and critical infrastructure and how disruption to critical infrastructure as a result of a catastrophe would impact logistics systems.
  - Correlate the prioritization of restoration of critical infrastructure with the suffering that can occur as a consequence.
  - Correlate the role that public health plays with minimizing the effects of a catastrophe, including the critical infrastructure needs for disaster health and medical (NRF ESF-8) response.

    - Contrast methods to effectively manage aid donations from national sources with international sources.
    - Contrast between preevent and postevent evacuation, and relocation.
    - Analyze major issues that may arise during a mass evacuation and resettlement, including possible methods of addressing these issues.
    - Analyze the relocation continuum (e.g., the different things/events that occur during relocation, such as emergency sheltering, temporary housing, and long-term housing and relocation postcatastrophe).
    - Appraise transportation modes for both evacuation and relocation.
- Outline of Topics
    - Chapter 5: Logistics
        - Overview
        - The Concept of *Convergence*
        - Logistics in Disasters Compared to Catastrophes
        - Challenges to Critical Resource Provision
        - Social and Cultural Context
    - Chapter 6: Critical Infrastructure/Key Resources (CI/KR)
        - Introduction
        - An Overview of Critical Infrastructure and Key Resources
        - The Effects of a Catastrophe on Infrastructure
        - The Critical Infrastructure Needs of Responders
        - Prioritization of Restoration of Critical Infrastructure
    - Chapter 7: Mass Care, Public Health
        - Introduction
        - Basic Vocabulary
        - Infectious Disease Vocabulary
        - Disease Control Mechanisms
        - Catastrophes and Public Health
        - Public Health Priorities
        - Infrastructure and Support Needed for Public Health
        - Role of Surge Capacity Planning in Catastrophes
        - Emergency Management: Public Health Collaboration in Catastrophes
    - Chapter 8: Mass Evacuation and Relocation
        - Introduction
        - Complexity and Causation
        - Defining Mass Relocation
        - Understanding Mass Relocation
        - Historic Perspective of Catastrophes and Mass Relocation

- Global Climate Changes and Mass Relocation
- Defining the Displaced
- Human Rights Dimensions of Mass Relocation
- Identifying the Potentially Displaced
- Mass Relocation as Mitigation
- Social Vulnerability
- Demographic Movement
- Continua of Displacement
- Displacement and Loss
- Involuntary Displacement and Recovery
- Resettlement
- Impoverishment Risks and Reconstruction Model
- Responsible Agencies in Mass Relocation
- Resettlement Action Plan
- The Near Future

• Section III Conclusion

CHAPTER 5

Logistics

5.1 LEARNING OBJECTIVES

As a result of this chapter, the reader should be able to:

- Compare current potential failure points (current and potential) in the national logistics system and resource mobilization with those created as a result of a catastrophe.
- Correlate the roles and limits of technology in catastrophe response.
- Correlate the nexus between logistic systems and critical infrastructure and how disruption to critical infrastructure impacts logistics systems.
- Contrast methods to effectively manage aid donations from national sources with international sources.

5.2 KEY TERMS AND PHRASES

- Convergence
- Critical resource provision
- Donations management

5.3 OVERVIEW

This chapter contains an overview of core issues related to critical resource provision following a catastrophic event. Emphasis is placed on material resources rather than information or personnel, although both are addressed. The reader is then provided with a short case study from the 2004 Indian Ocean tsunami and encouraged to explore the social and cultural considerations that may go into critical resource provision.

BOX 5.1 ROLE OF LOGISTICS IN CATASTROPHIC RESPONSE

- Get the right supplies and personnel to the place where they are needed at the correct time, as efficiently as possible.
- Make it possible for disaster responders to do their work.
- Make it possible for catastrophe-affected populations to acquire what they need for survival and rebuilding.

5.4 THE CONCEPT OF *CONVERGENCE*

This discussion provides an overview of the concept of convergence as it relates to critical resource provision in extreme events. The term *convergence* is frequently used in literature describing the social aspects of disasters and catastrophes as well as the mathematically informed study of logistics. In the study of emergency supply chains following extreme events, convergence takes on two distinct meanings.

Convergence, in the mathematical sense, refers to the coming together or approach toward a specific value or definitive point. In this case, you may be interested in determining the extent to which quantities of critical resource supplies converge or match up with the need for those same resources. This use of the term *convergence* is most often used by civil engineers who study logistic operations. Here, convergence is a goal of logistics management and one hopes to converge the two (need and supply) mathematical values to the same definitive point.

Those adopting a social science perspective may be interested in how best to coordinate this influx, when it is best to take measures to limit the flow, and the extent to which one type of convergence impacts another type of convergence (e.g., when those that are converging also bring material resources). Those adopting a mathematical perspective may be interested in critical supply optimization. In other words, how to foster an accelerated convergence between dynamic needs and supply flows of critical resources where the supply meets but does not greatly exceed the demand.

Critical resource provision after a catastrophic event involves people, material, and information from formal governments, national and local nonprofit groups, national and local private sector organizations, emergent groups, and the general public. The influx of people, material, and information is common and should be expected in a catastrophe, although local governments and organizations are rarely sufficiently prepared for it. One of the most important areas for jurisdictions to address is donations management. This influx can take the form of *external convergence* ("the notion of movement toward the disaster-struck

BOX 5.2 DEFINITIONS OF CONVERGENCE

1. *Convergence* in the social science sense refers to the influx of people, material, and information to sites associated with the disaster response environment.
2. According to Fritz and Mathewson (1957), *convergence* can include those individuals who come to help (both formal and spontaneous responders); those who return to the area (such as residents or business owners); those who anxiously look for loved ones; those who are curious about the event (including the general public or the media); and those who hope to exploit the situation to their advantage. In some responses, Kendra and Wachtendorf (2003) point out that we also may see the convergence of people who want to encourage or support the response efforts and those who have come to mourn or memorialize the victims. Material goods—both those that are gravely needed and those that are unnecessary and overburden the system—and information from official and unofficial sources also are likely to converge upon or at response-related sites. Here, convergence is a social/behavioral phenomenon whose positive and negative features must be contended with.

area from the outside") or *internal convergence* ("the movement toward specific points within a given disaster-related area or zone") (Fritz and Mathewson, 1957). Most often, both external and internal convergences occur. The greater the event—in terms of magnitude, scope, scale—as well as the greater the public attention to the event, the more likely it is that external convergence will take place. For catastrophes, external convergence will most likely occur.

5.5 LOGISTICS IN DISASTERS COMPARED TO CATASTROPHES

Quarantelli (2005) identifies the different characteristics of catastrophe. Each characteristic should be put in context with the potential impacts on the material supply chain. The reader is encouraged to read Wachtendorf et al. (2010), which adds to the characteristics of catastrophes and considers them in the supply chain context.

There are several ways in which catastrophes differ from disasters. These differences have implications for the logistical considerations of critical resource provision. Let us consider, in particular, material resources.

BOX 5.3 ISSUE: DAMAGE TO INFRASTRUCTURE

1. **Characteristics of catastrophe:** Most or all of the community-built structure is heavily impacted … [and] the facilities and operational bases of most emergency organizations are themselves usually hit.
2. **Consequence to the emergency supply chain:**
   - Damage to transportation infrastructure, such as roads and bridges, and port and airport facilities, creates difficulties navigating the impact zone, resulting in significant time delays in supply delivery.
   - Damage to communication infrastructure leads to inadequate communications.
   - Prepositioned supplies may be damaged.
   - Alternative space for supply warehousing and distribution may be difficult to identify.
   - The response and short-term recovery period may be protracted. A quick influx of long-term recovery-related supplies may unnecessarily clog the system.

BOX 5.4 ISSUE: MIGRATION ISSUES

1. **Characteristics of catastrophe:** Mass and extended out-migration of residents.
2. **Consequence to the critical supply chain:** Competing demands for emergency resources spread over a wider area, adding further complexity to the flow of goods as location of those in need changes.

BOX 5.5 ISSUE: LOCAL OFFICIALS ARE OVERBURDENED

1. **Characteristics of catastrophe:** Local officials are unable to undertake their usual work role, and this often extends into the recovery period.
2. **Consequence to the emergency supply chain:**
   - Overburden on the local system and difficulties contacting those who would normally communicate needs in a disaster.
   - Additional personnel from outside the area are especially needed for the emergency supply chain, even though integrating outside personnel can bring coordination challenges.

BOX 5.6 ISSUE: LOSS OF FUNCTIONALITY

1. **Characteristics of catastrophe:** Most, if not all, of the everyday community functions are sharply and concurrently interrupted.
2. **Consequence to the emergency supply chain:**
   - Some community-based organizations that would normally play a role in distribution (e.g., some local food banks or churches) are themselves significantly impacted or not allowed back into impacted areas.
   - Local commerce is impacted, making it more difficult to acquire large amounts of resources locally or to identify local suppliers still in operation.

BOX 5.7 ISSUE: INTENSIVE MEDIA SCRUTINY

1. **Characteristics of catastrophe:** The mass media system, especially in recent times, socially constructs catastrophes even more than they do disasters.
2. **Consequence to the emergency supply chain:**
   - Convergence of donated goods from across the county and around the world occurs on a larger scale.
   - As a result, logistics operators encounter additional challenges (more than the "typical disaster") in sorting perishable or low priority goods from high priority items.
   - Media may, moreso than in a disaster, serve a role in providing the first picture of critical needs, thereby impacting the flow of supplies through their particular framing of the event.

BOX 5.8 ISSUE: IMPORTANCE OF POLITICS

1. **Characteristics of catastrophe:** Because of the previous five processes, the political arena becomes even more important.
2. **Consequence to the critical supply chain:** Heightened discourse of who is at fault as different levels of government blame one another for inadequacies in supply distribution.

BOX 5.9 ISSUE: NO HELP FROM OUTSIDE

1. **Characteristics of catastrophe:** Help from nearby communities cannot be provided.
2. **Consequence to the emergency supply chain:**
   - Competition between communities for scarce resources rather than mutual aid provision
   - Greater need for external convergence equals:
     - Greater time delays in reaching impacted areas
     - Increased challenges in coordinating what is coming in

5.6 CHALLENGES TO CRITICAL RESOURCE PROVISION

We will now focus the discussion on the additional challenges to critical resource provision that may occur in a catastrophic event. Many of the logistical issues confronting communities impacted by a catastrophe are consistent with what is more typically categorized as a disaster. Although you may encounter this content in other readings, the relevance to catastrophes merits its inclusion.

Many of the same logistical issues that plague a large-scale disaster response emerge in responses to catastrophic events. It is worthwhile to consider these lessons learned from more routine or less catastrophic events since we have very limited exposure to catastrophes in modern times from which to draw. The magnitude of the requirements for an effective response correspondingly increases as the magnitude, scale, and scope of an event increases. Incident size places considerable strain on emergency supply chains and logistical systems. More organizations become involved, many from outside the area and with little experience working with each other in the past. A system that may prove fully sufficient for a disaster response may not come close to meeting the needs that emerge during a catastrophic event. Although large-scale events may effectively mobilize the general public, as such events capture the attention of national and global media, they also may result in a massive influx of unnecessary donations that create an environment where sorting, storing, and transporting donated material goods becomes extremely complex (Neal, 1994) (Figure 5.1).

Coordination of critical resources during catastrophic events requires a unified multiorganizational response. It is not unusual to see involvement from emergency management agencies (including all levels of government), other government agencies, community-based organizations, faith-based organizations, an umbrella organization called Voluntary Organizations Active in Disasters (VOADs), national nonprofit organizations, as well as their local chapters, emergent groups, and private sector organizations (at the national, regional, or local level).

It is essential for jurisdictions to be able to effectively mobilize, transport, inventory, track, store, and distribute material and human resources in a major catastrophe. Effective demobilization after an event is also crucial. In order to achieve this over-arching goal of effective logistical operations postcatastrophe, responses are best organized when there are preexisting plans for resource provision, prepositioned supplies, resilient communication infrastructure, and personnel are available with sufficient training in logistics management. Technology is also an important part of the solution when considering the problems related to critical

FIGURE 5.1 Donations mismanagement: Photos taken in the weeks following Hurricane Katrina and showing sites in Mississippi and Louisiana and the mishandling of donations from organizations outside the impact area. (Photos courtesy of the Disaster Research Center, University of Delaware.) These photos are from the weeks following Hurricane Katrina and include sites in Mississippi and Louisiana. They show the mishandling of donations from organizations outside the impacted area. (Photo courtesy of The Disaster Research Center, University of Delaware.)

resources provision, but technical solutions are not panaceas and should be considered alongside organizational and community solutions.

Preexisting memoranda of understanding and interagency agreements are useful organizing mechanisms to put action plans in place prior to an actual event. These allow for the speedy mobilization of many resources whose need can be anticipated following a catastrophe. Moreover, the prepositioning of supplies prior to an event can help to contend with immediate needs while additional postcatastrophe resources are transported to the region. At the same time, catastrophes are accompanied by many unpredictable circumstances. Supplies not previously anticipated may suddenly prove essential. Suppliers who are part of the preexisting arrangements may not be available to provide assistance at that time or may, themselves, be impacted. The need may exceed the amount that can be supplied through an existing contract. Prepositioned supplies such as medical caches may be destroyed in the incident or the extent of the postincident requirements may be underestimated. Anticipated methods of transportation may no longer be accessible. Any number of factors could still necessitate an adaptive and improvised logistics response.

If the communications infrastructure fails during an event, the ability to execute interagency coordination may become challenged. Without an adequate communications system in place, responders would experience difficulty in sharing information on need, supply availability, and supply delivery. After Hurricane Katrina, some officials in the hardest hit areas were forced to communicate back and forth using a helicopter, sending handwritten notes, operating as a new-age pony express. Officials had little way of knowing if requests were received and being processed, and some took measures into their own hands by commandeering supplies. The mass media also became a source of information dissemination during this event, with local residents and officials communicating their needs through the print and television media in the area. Although this latter strategy can prove somewhat effective, it generates delays, may not convey all community needs, the need can be misinterpreted, and may miss highly vulnerable areas where the media does not have a presence.

In a catastrophic event, understaffing of key response agencies and lack of training in logistics management may well generate substantial logistical problems for a response operation. Properly trained personnel are needed to oversee the transportation of goods, manage warehouses, complete contracts, and distribute goods. Fortunately, along with the prevalence of material convergence, the convergence of helpers is quite common in large disasters and catastrophic events (Fritz and Mathewson, 1957). Unfortunately, these helpers are both tightly and loosely organized, they represent both skilled and unskilled individuals, and they bring with them their own logistical burden (e.g., lodging, subsistence, sanitation considerations, etc.). It may be extremely difficult to differentiate the skilled from the unskilled helper. Any time devoted to logistical training of unskilled workers in the postcatastrophic environment takes critical time away from logistical response. Moreover, if volunteers only remain in the area for a short time, they take their newly acquired skills with them when they leave, resetting the learning curve each time new workers arrive that require training. The solution, however, should not be to ban all spontaneous volunteers. Not only is the objective highly unfeasible in the post-catastrophic setting, but loosely connected converging helpers bring skills and knowledge lacking in those with formal organizational ties. If a response system is too closed and resistant to volunteer efforts, important and much-needed skills may be overlooked. These issues all bolster the need for jurisdictions to have good plans for the use of volunteers including vetting, training, credentialing, etc.

Sites are needed to stage goods coming in from outside areas; to sort, inventory, store, and move to other facilities the resources that have arrived; and to distribute donated and other goods to those in need and assistance organizations. Sites are also needed to receive, credential, assign, and deploy spontaneous volunteers. Some sites function within the formal response system, while others operate independently, even as somewhat of an underground operation.

Research has even pointed to the "disaster recovery underground" that emerges in the weeks after a large-scale event (Ruback, forthcoming), where those citizens who are able to tap into an information-sharing network of new (often short-lived) donation sites can often secure the best resources. Of course, the presence of these emergent supply sites relies on some level of community function. In catastrophic events, where transportation systems and other critical infrastructure are decimated, it may take significantly longer for these formal and informal assistance sites to become established, or they may emerge in areas further afield from the impact zone and closer to where evacuees have temporarily reestablished themselves.

Technology can be an important tool for catastrophic logistical operations. Clearly, more advanced communication systems (such as satellite technology) can enable information exchange even when other ground-based systems (like hard-wire and cellular telephone service) fail. Improved and widely used logistical software as well as the use of Global Positioning Systems (GPS) and Geographic Information Systems (GIS) technologies could better enable resources tracking, destination navigation, and visibility of inventory. However, solely relying on technical solutions is not the answer to the critical resource provision challenge. According to Cannon (2000), a technological approach to solving disasters and, by extension catastrophes, creates a situation where institutions conceive of problems in terms of what their own capacities are. That is, solutions are defined in terms of what seems *possible* rather than what is really *needed*. Technology alone is unlikely to solve issues related to increased social vulnerability, for example. It may not be able to anticipate every conceivable need to minimize system breakdown. It may not be updated as poorer communities opt to invest public funds elsewhere, or where even affluent communities consider emergency planning a low priority. The technology may not be standardized and may fail to operate, due to damage to the supporting infrastructure and/or a lack of interoperability among systems, with neighboring jurisdictions or with different levels of government. Lack of interoperability between logistics systems was a significant obstacle in the Hurricane Katrina response (Holguin-Veras et al., 2007). The same is true, even more so, in a catastrophic event. Technology should be viewed as a useful tool in logistics management, but managing critical resources in a catastrophic environment must also involve organizational solutions that facilitate building on local knowledge, adaptive solutions, and organizational learning as the event unfolds.

At the time of this writing, the U.S. Department of Homeland Security was operating under the 2008 National Response Framework. The specifics of plans, procedures, and protocol regularly change; specific components of the framework are not outlined in this chapter, but are discussed in greater detail in Section IV. Instead, what we will discuss here are general principles pointed to in the Emergency Support Function #7—Logistics Management and Resource Support Annex (http://www.fema.gov/pdf/emergency/nrf/nrf-esf-07.pdf). The reader is encouraged to refer to the most recent version of this national document as well as the logistics section or annex of their own State Emergency Management Plan.

5.7 SOCIAL AND CULTURAL CONTEXT

We will now examine the impact social and cultural context has in determining the appropriateness of critical supplies provision. We will use the December 2004 tsunami as a case study to explore this issue further (Figure 5.2a and b). This sets the stage for our discussion of social and cultural appropriateness in relief provision.

A field team of researchers from the Disaster Research Center (University of Delaware) and the University of North Texas took these photos following the 2004 Indian Ocean tsunami. This team was sponsored by the Earthquake Engineering Research Institute, the National Science Foundation, and the Sea Grant Program at the

(a)

(b)

FIGURE 5.2 Tsunami temporary shelter: These photos were taken in India following the 2004 Indian Ocean tsunami and show temporary shelters. (Photos courtesy of the Disaster Research Center, University of Delaware.)

University of Puerto Rico/Mayaguez to examine the social impacts of this event caused by the December 26, 2004, Great Sumatra–Andaman earthquake.

On December 26, 2004, a large earthquake occurred off the west coast of Sumatra, Indonesia. The great Sumatra earthquake had a magnitude of between 9.1 and 9.3 and generated a tsunami that killed 225,000 people. Most coastal nations along the Indian Ocean were impacted, with Indonesia, Sri Lanka, India, and Thailand suffering the most damage and the highest number of fatalities. By all accounts, this was a catastrophic event.

Following the tsunami, impacted countries saw a tremendous influx of assistance from abroad. These photos (Figure 5.2a and b) were taken just outside a fishing village in the Indian state of Tamil Nadu. Residents of this village were camping near an earthen elevated highway because of the devastation to built structures in their community and their fear of another tsunami. Residents stated that the elevation of the highway, only a short run from the temporary camp, would provide some protection should another tsunami occur.

In the two photos, we see two types of temporary shelters. On one side of the road to the village were ad hoc shelters constructed out of locally available tarps, blankets, thatch, and poles (see Figure 5.2a). On the other side of the road were well-constructed camping tents donated by an international nonprofit organization (see Figure 5.2b).

Many of the same lessons learned from the research on disaster events apply to catastrophic scenarios. It is not uncommon to hear of winter clothing donated to hot climates, baby formula donated to countries accustomed to nursing, and meat-based products donated to primarily vegetarian countries. As the previous photos from Hurricane Katrina show, sometimes donated goods are simply dumped on the side of the road by hired drivers who refuse to wait for extended periods of time to be unloaded at a receiving facility.

While the need to provide assistance may be great in a catastrophe, we should not ignore the social and cultural settings in which relief is provided. The contrast between these settings and those with which we are most familiar may be most stark when providing international assistance. At the same time, communities are diverse and there may be unique considerations even within the communities in which we live. Enclaves of immigrant populations can present challenges ranging from the ability to communicate both verbally in writing and cultural barriers that include dietary needs as well as fear or mistrust of persons in authority.

If the goal of humanitarian assistance is truly to help those in need, it is not enough to assume that any help is better than no help at all. If we want our assistance to be meaningful and truly helpful, it is worthwhile to mobilize resources that take social norms, values, and local conditions into account. This is why it is so critical to work with those organizations that have a long history of involvement in the impacted communities and who understand local issues.

5.8 DISCUSSION QUESTIONS

1. Brainstorm the kinds of material items you would need to consider were you to serve as logistics coordinator for a catastrophic event. Include resources related to: subsistence, energy (e.g., oil and electricity); administrative supplies; petroleum products; engineering and construction material; personal demand products (e.g., water and ice); major items (e.g., mobile units); medical material, property (e.g., space); facilities; telecommunications and support; or transportation.
2. Identify five challenges that may prevent the timely delivery of essential material following a catastrophe.
3. Imagine that you are a relief worker assigned to deliver the donated tents to persons made homeless from a catastrophe. What questions would they ask of the community? What actions would you take in the delivery of these temporary shelters?

REFERENCES

Cannon, T. 2000. Vulnerability analysis and disasters. In *Floods*, ed. D.J. Parker, 45–55. London: Routledge.

Fritz, C. and J.H. Mathewson. 1957. Convergent behavior: A disaster control problem. Special Report for the Committee on Disaster Studies. Washington, D.C.: National Academy of Sciences.

Holguín-Veras, J., N. Pérez, S. Ukkusuri, T. Wachtendorf, and B. Brown. 2007. Emergency logistics issues impacting the response to Hurricane Katrina: A synthesis and preliminary suggestion for improvement, *Transportation Research Record* 2022, 76–82.

Kendra, J.M., and T. Wachtendorf. 2003. Reconsidering convergence and converger legitimacy in response to the World Trade Center disaster. Terrorism and disaster: New threats, new ideas, ed. L. Clarke. *Research in Social Problems and Public Policy* 11: 97–122.

Neal, D.M. 1994. The consequences of excessive unrequested donations: The case of Hurricane Andrew. *Disaster Management* 6: 23–28.

Quarantelli, E.L. 2005. Catastrophes are different from disasters: Implications for crisis planning and managing drawn from Katrina. Understanding Katrina: Perspectives from the Social Sciences. Social Science Research Council, Brooklyn, NY. Online at: http://understandingkatrina.ssrc.org (accessed January 10, 2010).

Ruback, J.R. (forthcoming). *Hands-on disaster recovery: A citizen's guide.*

Wachtendorf, T., B. Brown, J. Holguin-Veras, and S. Ukkusuri. 2010. Catastrophe characteristics and the impact on critical supply chains: Material convergence following Hurricane Katrina. University of Delaware. Disaster Research Center. Preliminary Paper 365.

CHAPTER 6

Critical Infrastructure/ Key Resources (CI/KR)

6.1 LEARNING OBJECTIVES

As a result of this chapter, the reader should be able to:

- Correlate the roles and limits of technology in catastrophe response.
- Correlate the nexus between logistic systems and critical infrastructure and how disruption to critical infrastructure as a result of a catastrophe would impact logistics systems.
- Correlate the prioritization of restoration of critical infrastructure when the suffering can occur as a consequence.

6.2 KEY TERMS AND PHRASES

- Critical infrastructure and key resources (CI/KR)
- CI/KR protection
- CI/KR restoration
- CI/KR sectors
- CI/KR stakeholder
- Government Coordinating Council (GCC)
- Homeland Security Presidential Directive 7 (HSPD-7), Critical Infrastructure Identification, Prioritization, and Protection
- National Infrastructure Protection Plan (NIPP)
- Sector Coordinating Council (SCC)
- Sector Specific Agency (SSA)
- Sector Specific Plan

6.3 INTRODUCTION

Critical infrastructure and key resources (CI/KR) protection is a highly complex and involved topic, the subject of myriad publications. This discussion *is not* intended to impart expert knowledge of CI/KR and/or CI/KR protection. Rather, its purpose is to provide a general overview of CI/KR, CI/KR protection so the reader can relate this to the issues of interruption and restoration of CI/KR after a catastrophe.

It is important to note that much of the existing material on CI/KR is dedicated to CI/KR protection (e.g., prevention) rather than restoration. Furthermore, the body of work on CI/KR, with few exceptions, does not focus on the impact of catastrophic disasters so much as on disasters that do not rise to the catastrophic level. Furthermore, since we have very limited experience with catastrophic disasters in the United States, we also have little experience with regard to how a catastrophe will impact CI/KR. However, we do have ample experience with lower level events and how they impact CI/KR. We also have limited useful information on how catastrophes have impacted CI/KR in other industrialized countries. Therefore, it may be possible to project the plausible effects on CI/KR from a catastrophe within the United States.

6.4 OVERVIEW OF CRITICAL INFRASTRUCTURE AND KEY RESOURCES (CI/KR)

In this discussion, the reader will be presented with an overview of CI/KR and CI/KR protection. Additionally, the reader should appreciate that both CI/KR protection and restoration are shared responsibilities for which federal, state, and local government entities must work closely and coordinate with the private sector, since approximately 85% of the critical infrastructure in the United States is privately owned and operated. Therefore, the process of protecting CI/KR and the process of restoring CI/KR postdisaster or catastrophe is not always seamless because different stakeholders can have varying approaches or priorities. Note that throughout this chapter, the term *state and local government* also includes territorial and tribal governments.

6.4.1 What Is CI/KR?

The official definition of critical infrastructure is set forth in the United States Patriot Act (P.L. 107-56), which defines CI as:

> Systems and assets, whether physical or virtual, so vital to the United States that the incapacity or destruction of such systems and assets would have a debilitating impact on security, national economic security, national public health or safety, or any combination of those matters.

As this is the definition as prescribed by law, federal government agencies (e.g., Department of Homeland Security (DHS)) either use this definition or similar ones that vary by only a few words, as in the DHS's National Infrastructure Protection Plan (NIPP). The definition of CI is fairly straightforward, and should not need further definition.

Key resources are defined in both the Homeland Security Act of 2002 (P.L. 107-296) and the NIPP as:

> Key resources are publicly or privately controlled resources essential to the minimal operations of the economy and government.

As noted by the Government Accountability Office, the Homeland Security Act does not articulate exactly what key resources are, but it does view "key resources as distinct from critical infrastructure, albeit worthy of the same protection."

The NIPP further defines key resources by equating key resources with the earlier used phrase key assets, which are defined as:

> Individual targets whose destruction would not endanger vital systems, but could create local disaster or profoundly damage the nation's morale or confidence.

The NIPP clearly states that *key assets* and *key resources* are one and the same and that key resources is now the commonly used and correct term. Thus, key resources can essentially be seen as the component parts that comprise critical infrastructure.

6.5 CI/KR PROTECTION

The basics of CI/KR protection are set forth in Homeland Security Presidential Directive 7 (HSPD-7), Critical Infrastructure Identification, Prioritization, and Protection. The primary effect of HSPD-7, which concerns this discussion, is that HSPD-7 establishes the basic framework for protection of CI/KR, sets forth the responsibilities of DHS and other federal agencies with respect to protection of CI/KR, and identifies 17 distinct areas of the national economy as CI/KR. These sectors are:

BOX 6.1 CI/KR AREAS OF THE NATIONAL ECONOMY

Agriculture and Food
Chemical
Communications
Water
Emergency Services
Government Facilities
National Monuments and Icons
Postal and Shipping
Transportation Systems
Banking and Finance
Commercial Facilities
Dams
Defense Industrial Base
Energy
Information Technology
Nuclear Reactors, Materials, and Waste
Public Health and Healthcare

HSPD-7 also specifically assigns primary responsibility for each CI/KR sector to one or more federal agencies, each of which is called a Sector Specific Agency (SSA). The sector specific agencies responsible for each sector are:

BOX 6.2 SECTOR SPECIFIC AGENCIES RESPONSIBLE FOR EACH SECTOR

Sector	Sector Specific Agency(ies)
Agriculture and Food	Department of Agriculture (USDA): meat, poultry, and eggs Department of Health and Human Services, Food and Drug Administration (FDA): all other food and agriculture
Banking and Finance	Department of the Treasury
Chemical	Department of Homeland Security
Commercial Facilities	Department of Homeland Security
Communications	Department of Homeland Security
Dams	Department of Homeland Security
Defense Industrial Base	Department of Defense
Emergency Services	Department of Homeland Security
Energy	Department of Energy (except for commercial nuclear power facilities and nuclear waste)
Government Facilities	Department of Homeland Security
Information Technology	Department of Homeland Security

(continued on next page)

Sector	Sector Specific Agency(ies)
National Monuments and Icons	Department of the Interior
Nuclear Reactors, Materials, and Waste	Department of Homeland Security
Postal and Shipping	Department of Homeland Security
Public Health and Healthcare	Department of Health and Human Services
Transportation Systems	Department of Homeland Security
Water	Environmental Protection Agency

Each SSA is responsible for leading efforts to protect that agency's particular CI/KR sectors. This responsibility includes overseeing development of a Sector Specific Plan (SSP) that explains how the NIPP will be implemented for the specific sector (e.g., how the sector will be protected, given the guidelines and requirements set forth in the NIPP and the unique characteristics of the particular sector). The SSAs also coordinate private industry and government representatives who are involved in the specific sector.

Private industry representatives for the sector meet as members of the Sector Coordinating Council (SCC), which is intended to have broad representation of stakeholders of the particular CI/KR sector, including owners, operators, and industry associations. The SCC is also intended to serve as a method for communication between private industry representatives as well as to be the primary means for the government and the sector's private industry representatives to coordinate and communicate concerning CI/KR protection.

Representatives of federal, state, and local government agencies meet as members of the Government Coordinating Council (GCC) and provide interagency coordination for CI/KR protection. The GCCs are the government counterparts of the SCCs, and the GCCs facilitate interagency and cross-jurisdictional coordination and communication.

6.6 EFFECTS OF A CATASTROPHE ON INFRASTRUCTURE

All CI/KR sectors have weaknesses and are vulnerable to damage, disruption, and destruction, and any catastrophic disaster that can cause physical damage to an area (such as an earthquake) will have a physical impact on the built environment and, hence, on CI/KR. However, as with a lower level event, the effects on CI/KR of a catastrophe can vary depending on the specifics of the catastrophe, including the type of event, location, magnitude, and the time of year (time of day) when the event occurs as well as specific characteristics and vulnerabilities of CI/KR in the impacted areas.

For example, while catastrophic earthquakes and hurricanes can both cause massive and widespread damage to CI/KR, their effects on CI/KR may differ from each other. Equally important is that the same type of catastrophic event may have a significantly different impact on CI/KR in different parts of the country (e.g., an earthquake in the Central United States versus the West Coast).

On the other hand, though a pandemic is potentially much more catastrophic in terms of loss of human life than an earthquake or hurricane due to its worldwide impact on human beings, it may cause little or no direct damage to the physical infrastructure. However, attrition of critical personnel who are responsible for the operation of CI/KR, brought about by a pandemic, can cause significant disruption of CI/KR. This also can lead to delaying of routine maintenance and emergent repairs, all of which can bring about physical damage to the infrastructure and cascading failure of other sectors.

Perhaps the best way to predict the impact of a catastrophe on CI/KR is to consider the damage estimates for two potential catastrophe events: a South Florida major hurricane scenario, known as Hurricane Ono, and a New Madrid Seismic Zone (NMSZ) large-scale earthquake.

6.6.1 South Florida Hurricane

In the South Florida scenario, Hurricane Ono makes landfall just north of Fort Lauderdale as a catastrophic category 5 hurricane with sustained winds in excess of 155 miles-per-hour. The storm remains at category 4 strength as it dumps up to 15 inches of rain on South Florida and grazes Lake Okeechobee. The combination of rainfall, storm surge on the lake, and tornados spawned by the hurricane result in damages to the Lake Okeechobee flood control system, including breaches of three separate sections of the Herbert Hoover Dike, while the storm also causes the failure of a flood control structure on the St. Lucie Canal. The resulting damage sends millions of gallons of water from the lake south toward the coastal populations centers from West Palm Beach to Miami. After 36 hours over South Florida, Ono exits the west coast of Florida into the Gulf of Mexico at Pinellas County as a category 2 storm. Back over warm water of the Gulf, Ono rapidly reintensifies, turns north, and makes a second landfall between Mobile, Alabama, and Pensacola, Florida, as a category 4 hurricane.

In the course of less than two days, Ono floods most of South Florida with 1 to 4+ feet of water, which will remain for weeks; destroys or severely damages more than 750,000 buildings; forces almost 3 million people to evacuate before the storm and almost 800,000 people to use emergency shelters; renders the homes of more than 3.8 million people uninhabitable, and leaves more than 6 million people without electricity for weeks. Ono also devastates Florida's transportation infrastructure as well as the state's agriculture, service, and tourism industries. Ono results in damage to all of the CI/KR sectors and major damage to or disruption of at least 12 of the 17 CI/KR sectors. Because of where the storm hits, the Nuclear Reactor, Material, and Waste, Chemical, Information Technology, Banking and Finance, and Defense Industrial sectors may escape major damage.

6.6.2 New Madrid Earthquake

In the NMSZ earthquake catastrophic scenario, a magnitude 7.7 earthquake on the New Madrid Fault in northeastern Arkansas results in massive damage to parts of at least five states, and the earthquake is felt over much of the eastern half of the United States.

In the course of just a few seconds, the earthquake kills over 3,000 people, injures over 60,000, damages or destroys almost 750,000 buildings, and leaves millions of people in the immediate impact area without power, water, sewer, and other necessary services. The earthquake severely damages or destroys significant portions of the critical infrastructure in the impact zone, including transportation systems, electrical power systems, gas, petroleum and other pipelines, and communications infrastructure. The damage to electrical power transmission systems causes a "cascading catastrophic failure" of the eastern power grid, which interrupts power to much of the eastern third of the United States. Electrical interruptions in areas some distance from the epicenter are short-lived, lasting only hours to days; however, closer to the epicenter, it may take several months to restore electrical service to undamaged buildings and CI/KR. Damage to or destruction of gas and petroleum transmission lines results in persistent shortages of gas and petroleum products throughout much of the Northeast and Midwest. In addition, the collapse of bridges over the Mississippi River blocks the transportation of Midwest farm products and many industrial goods, both of which constitute major sources of export earnings.

The NMSZ earthquake results in some damage to all of the CI/KR sectors and major damage or disruption of at least 13 of the 17 sectors. Because of facility locations, it is possible that the Nuclear Reactors, Material, and Waste, Banking and Finance, and Defense Industrial sectors may avoid major damage.

Both of these scenarios will result in significant damage to and interruption of essentially every sector of critical infrastructure, albeit different parts of each sector. It is clear that catastrophic disasters that have a physical impact have the potential to severely affect most or all of the CI/KR sectors for extended periods of time.

6.7 CRITICAL INFRASTRUCTURE NEEDS OF RESPONDERS

In order to effectively respond to a catastrophe, certain CI/KR needs will have to be met. If not, either by the existing (undamaged) infrastructure or temporary/replacement infrastructure, responders will not be able to effectively carry out their missions, and the response and recovery will suffer and be delayed.

While all CI/KR sectors are critical to the nation, only some are *immediately* necessary to support responders to a catastrophe. Those CI/KR sectors that can be considered necessary for response include:

- Agriculture and food
- Communications
- Energy (particularly electric power and fuel)
- Information technology
- Transportation systems
- Water

While services or products of these sectors are necessary for responders, they may not all be immediately necessary for response to begin. This is because many, if not all responders, will most probably bring resources to meet some of their initial essential CI/KR needs. This is particularly true of the federal response teams with regard to food, communications, energy, information technology, and water.

The one CI/KR sector that is immediately essential to all responders, including units that are designed to be self-sufficient during the initial few days of response (e.g., Disaster Medical Assistance Teams (DMAT) and Urban Search and Rescue (USAR) teams) is transportation systems. Without some type of transportation, responders cannot actually travel to and then in and around the impact area. Furthermore, even if teams are initially self-sufficient, they will eventually consume the supplies that they brought with them and will require resupply, often in as little as three days. Therefore, it is critical that advance plans be made for resupplying the critical needs of responders.

It is also critical to note that, because of the massive scope of a catastrophe, there will be many more responders to a catastrophic disaster than to a disaster, hence, significantly greater demands placed on the existing infrastructure as well as a need for significantly greater resupply levels. It also is likely that the external response teams will be on the ground for a much longer period of time following a catastrophe as compared to a disaster, thus further straining the resupply logistics both in terms of quantity and duration at a time that transportation resources are likely to be minimal. While there is no easy way to estimate the total number of responders to a given event, a quick look at some of the planning work for Hurricane Ono shows that, from a search and rescue (SAR) standpoint alone, there will be a need for thousands of first responders (in fact, many more than are readily available in the entire country); therefore, it would be prudent to plan for massive numbers of responders.

Lastly, it is critical to remember that responders will not be the only ones with CI/KR needs. There also will be a large number of survivors of the catastrophe and they will also place demands on the infrastructure, particularly due to their need for food, water, and shelter. Briefly looking at the Hurricane Ono scenario again shows just how great the magnitude of this issue can be. The projections for South Florida are that there will be approximately 4.3 million survivors left in the area after Ono. The vast majority of these people will have no electricity, most will not have sufficient shelter, and many will have insufficient supplies. Therefore, these survivors will require life-sustaining food, water, healthcare, and shelter. The task of providing three hot meals per day for survivors (one of Florida's goals) will be an incredibly daunting task given the severe damage to CI/KR sectors and the massive logistical requirements for meeting this goal.

6.8 PRIORITIZATION OF RESTORATION OF CRITICAL INFRASTRUCTURE

Any treatment of CI/KR issues postcatastrophe also has to deal with restoration of the various damaged CI/KR sectors. Unfortunately, in the aftermath of a catastrophe, there will not be sufficient resources to restore all sectors at the same time. Restoration, therefore, will require triage (e.g., restoration of some CI/KR life-sustaining sectors will have to be prioritized while restoration of other sectors will suffer as a result). As with protection and response, different government agencies, private sector entities, and individuals will have different ideas of what should be restored first, and, in the absence of advanced planning, lack of uniform agendas may unfortunately become the norm. Therefore, preevent CI/KR restoration planning is essential to ensure that the most good is done for the most survivors in the shortest amount of time. Now is a good time to remember our discussion concerning ethics in Chapter 3.

It may be easy to quickly determine that certain CI/KR life-sustaining sectors require immediate attention for restoration (e.g., Agriculture and Food, Emergency Services, and Water). It also may be easy to decide that restoration of others can wait for some time (e.g., National Monuments and Icons), other than for emergent tasks necessary to prevent further damage. However, the vast majority of CI/KR structures do not readily lend themselves to rapid determination of restoration priority. Can the Postal and Shipping sector be delayed for some time or is it needed to deliver life-sustaining material from other essential sectors? Can we forego restoration of the Energy sector for several weeks, say the NMSZ earthquake occurs in the dead of winter? If we allow restoration of commercial facilities to be significantly delayed, how many businesses will survive? Without rapid restoration of CI/KR, the impacted area begins to spiral out of control, with businesses failing or at least leaving the area and people relocating permanently in search of stable employment. In fact, one of the avowed goals for the South Florida Catastrophic Plan is rapid restoration of as many CI/KR sectors as possible to prevent mass relocation of the population who are seeking economic opportunity elsewhere. Sadly though, there may be little that can be done to avoid such an eventuality following a catastrophe where CI/KR is severely impacted.

6.9 DISCUSSION QUESTIONS

1. Think about the potential effects of one or more of the following catastrophes on critical infrastructure:
    - A 7.2 magnitude New Madrid Earthquake
    - Landfall of a major hurricane (e.g., category 4 or 5)
    - Pandemic
    - Major volcanic eruption of Mt. Rainier in the Pacific Northwest
2. Based on your answers to the question above, explore what might be the priority for restoration of CI/KR sectors with regard to the same catastrophic events and time of occurrence:
3. Is the division of several CI/KR sectors between multiple SSAs problematic, either from the standpoint of CI/KR protection or restoration postdisaster? In particular, the Agriculture and Food Sector is divided between the USDA and the FDA, while the Department of Energy has responsibility for the entire energy sector except for nuclear power, which is the responsibility of the Nuclear Regulatory Commission.
4. Would multiple SSAs for a single sector give rise to the potential for disconnect in planning for restoration of services post catastrophe?

FURTHER READINGS

Department of Homeland Security. 2008. Critical infrastructure and key resources. Washington, D.C.: DHS. Online at: http://www.dhs.gov/xprevprot/programs/gc_1189168948944.shtm (accessed January 10, 2010).

Department of Homeland Security. 2006. National infrastructure protection plan. Washington, D.C.: U.S. Dept. of Homeland Security. Online at: http://www.dhs.gov/xlibrary/assets/NIPP_Plan.pdf (accessed January 10, 2010).

Federal Emergency Management Agency (FEMA). 2008. Federal interim contingency plan—Predecisional draft: New Madrid Seismic Zone catastrophic earthquake response planning project. Washington, D.C.: FEMA.

Florida Division of Emergency Management, Federal Emergency Management Agency. 2008. Hurricane Ono consequence projections, full. Washington, D.C.: FEMA.

Florida Division of Emergency Management, Federal Emergency Management Agency. 2007. Florida catastrophic planning—Key planning assumptions. Washington, D.C.: FEMA.

Freeman, P.K. 2000. Infrastructure, natural disasters and poverty. In *Managing disaster risk in emerging economies*, eds. A. Kreimer and M. Arnold. Washington D.C.: World Bank Publications.

George W. Bush Administration. 2003. National strategy for the physical protection of critical infrastructures and key assets. Washington, D.C.: The White House.

Government Accountability Office, Larence E.R. 2007. GAO Report 07-1075-T Critical infrastructure—Sector plans complete and sector councils evolving. Washington, D.C.: GAO. Online at: http://www.gao.gov/new.items/d071075t.pdf (accessed January 10, 2010).

Murray, A.T., and T.H. Grubesic. 2007. *Critical infrastructure: Reliability and vulnerability*. Berlin: Springer-Verlag.

Radvanovsky, R. 2006. *Critical infrastructure: Homeland security and emergency preparedness*. Boca Raton, FL: Taylor & Francis Group.

U.S. Congress. House. Select Bipartisan Committee to Investigate the Preparation for the Response to Hurricane Katrina. 2006. *A failure of initiative*. 109th Cong., 2d S. Online at: http://www.gpoaccess.gov/katrinareport/mainreport.pdf (accessed January 10, 2010).

CHAPTER 7

Mass Care
Public Health

7.1 LEARNING OBJECTIVES

As a result of this chapter, the reader should be able to:

- Correlate the role that public health plays with minimizing the effects of a catastrophe including the critical infrastructure needs for disaster health and medical (ESF-8) response.

7.2 KEY TERMS AND PHRASES

- Centers for Disease Control and Prevention (CDC)
- Disease control mechanisms
- Emergency Support Function (ESF) #8: Health and Medical Services
- Endemic
- Epidemic
- Epidemiology
- Etiology
- Herd immunity
- Immunity
- Infectious disease
- Isolation
- Mass relocation
- National Disaster Medical System (NDMS)
- Quarantine
- Pandemic
- Public health

- Social distancing
- Surge capacity
- Surveillance
- U.S. Department of Health and Human Services (HHS), Office of Preparedness and Response
- Vaccination
- Vector
- Vector control

7.3 INTRODUCTION

This discussion is not intended to make readers literate in public health, but rather public health aware, so that they can fruitfully interact with the skills and needs of public health practitioners in the catastrophe context. The author is not a health professional, so don't be concerned about the subject content as it is mainly aimed at emergency management practitioners.

7.3.1 Role of Public Health

In all catastrophes, whether or not the incident is caused by a health phenomenon like a pandemic, the event will cause significant threat or actual harm to the affected population at the very time that people need their physical, mental, and emotional capabilities to be at their very best to deal with losses and move productively toward stabilization and some kind of recovery. The role of public health in this context must not be underestimated; it plays a direct role in helping people survive and recover from physical and mental trauma, and a crucial role in helping both individuals and communities marshal their resources for the daunting tasks ahead in reconstruction and recovery (Noji, 1997). Public health also plays an important role in protecting the health of emergency response personnel (Landesman, 2001). The reader would do well to consider public health issues throughout the remaining discussions in this book, so as to be able to start integrating the public health viewpoint into emergency management thinking with respect to catastrophic planning and response.

7.4 BASIC VOCABULARY

Public health is a broad interdisciplinary field of applied science, using some of its own methods and tools as well as borrowing concepts and methods from medicine, biology, sociology, psychology, anthropology, statistics, and ethics. Note that public health focuses on populations, while medicine typically focuses on individuals.

While the field of *epidemiology* can deal with epidemics, it is much broader and is intrinsic to virtually all aspects of public health. It is the discipline that combines the use of biomedical, human behavior, and statistical tools to investigate, describe, and analyze population-based health phenomena (Friis and Sellers, 2004). Epidemiologists are intimately involved in discovering the pathways and extent of infectious disease outbreaks, but are equally important in investigating all kinds of other disorders, including chronic diseases, injuries, auto-immune diseases, mental health problems, and even such behavioral problems as homicide and suicide. Epidemiologists also play an important role in helping scientists test new medical and pharmacological treatments for effectiveness in a trial population.

BOX 7.1 AN EPIDEMIOLOGIST IS TO ...

An *epidemiologist* is to public health what a police detective is to law enforcement.

Surveillance is one of the tools used by epidemiologists to analyze how health phenomena come about (e.g., injuries, sexually transmitted diseases) and are propagated. For public health officials in a catastrophic situation, surveillance is also one of the key tools used to determine whether the emergency response is having its intended effect of decreasing symptoms or needs. For an excellent description of this, see Sundnes and Birnbaum (2003).

BOX 7.2 KEY POINT

Public health is the "care" of the entire population, not just a single patient. It involves prevention and control of infectious diseases, injuries and other community-wide health problems.

7.4.1 Public Health versus Medicine

You may have difficulty comprehending how public health and medicine articulate with each other. There are some similarities with the articulation between emergency management and front-line emergency response practitioners. Both public health and emergency management are oriented toward the well-being of the entire affected population, while medicine and "rescue" personnel typically channel their efforts toward one person at a time. People who work at the individual level, such as medicine or first response, often staff the ranks of both public health and emergency management organizations. In both cases, there are staffers who practice both at the public and individual levels.

Also note that medicine and first response tend to be more technology-oriented, while public health and emergency management are more organizational and use knowledge of science and management methods to make sure resources are applied where they might do the most good. Finally, both public health and emergency management have a strong history of trying to prevent and mitigate situations before they evolve into emergencies or something worse. While medical and first response personnel do participate in prevention activities, they are most often engaged in trying to ameliorate problems that have already exerted themselves.

7.5 INFECTIOUS DISEASE VOCABULARY

> **BOX 7.3 VECTORS ARE ...**
>
> Vectors are the transmitters of disease-causing organisms that carry the pathogens from one host to another.

The word *etiology* is one that emergency managers will hear public health workers use frequently, and is a very useful word that should be incorporated into emergency management. Etiology means the entire causal chain and pathway of a disease or health condition. For example, the etiology of malaria includes a microbial blood parasite that is transmitted from human to human (or animal to human, in some cases) via a particular kind of mosquito. When a mosquito bites a human or other primate that carries the parasite that causes malaria, the blood the mosquito ingests brings the parasite into the mosquito's digestive tract. The parasite has to mature in the gut of the mosquito before it is capable of infecting another human. The mosquito, in turn, lives only in certain temperature and humidity conditions. Once the infected female mosquito bites a human, she transmits the parasites and that human becomes a new "host" for the parasite. The mosquito is termed a "vector" because it transmits the disease. Armed with knowledge of the etiology of malaria, public health personnel can strategize points in the process where an intervention can take place to stop disease transmission. For example, swamp drainage or the use of insecticides can minimize the vector (mosquito) population, and the use of window screens or bed netting can decrease the probability that mosquitoes can bite new victims. In a similar way, an emergency manager who knows the etiology of floods in a particular community or geography can strategize methods of mitigation to avoid future flood damages in areas known to flood.

It is important that emergency managers understand the three "emic" words. *Endemic* is when a disease just exists in a population at a baseline level, *epidemic* is when an unusually large number of cases occurs in a group of people (as small as a single school, or as big as a whole country), *pandemic* is when an epidemic spreads throughout the world. Public health personnel may use these words casually, expecting that their emergency management colleagues will understand them. For more explanation of these terms, see Friis and Sellers (2004).

Immunity is when an individual can no longer become infected with a specific organism. There are some infectious organisms to which humans cannot become immune, e.g., malaria. For other microbes, immunity can occur naturally because of surviving a prior infection, or by vaccines for specific illnesses, e.g., measles. The concept of *herd immunity* is important to understand, as it plays an important role in strategies to control new or reintroduced diseases, and helps emergency managers understand why it is not necessary to inoculate everyone in a population when supplies are limited. To learn more about this, see Friis and Sellers (2004), as previously mentioned. The Centers for Disease Control and Prevention (CDC) Web site (www.cdc.gov) also provides a good description of herd immunity and how it is used to control the spread of diseases in a population in which 100% vaccination cannot be achieved. In short, the concept is taken from veterinary medicine, in which it was found that one could decrease the probability of disease spread in a herd of domestic animals by vaccinating a significant percentage of the herd without having to vaccinate all of the animals. In order for disease-causing microbes to spread in a population, they need to pass from an infected individual to others that are vulnerable to the microbe. If one vaccinates a majority of the animals in the herd, the probability increases that the microbes will not find sufficient vulnerable individuals to spread widely, and the infection stops.

Social distancing, quarantine, and isolation are all strategies for decreasing exposure to microbes (viruses, bacteria, etc.) that are passed from one person to the next. Emergency managers may become involved in preparation for and enforcement of quarantine and isolation orders because they require facilities, food delivery, and, potentially, law enforcement. Note that social distancing has taken on new meanings recently to include increasing the distance between individuals when talking (at least three to six feet), the use of N95 disposable particulate respirators or surgical masks on people who are ill or suspected of being contagious, and the use of numerous strategies to maintain hand sanitation. Other forms of social distancing that have proved successful in the past include such strategies as canceling or prohibiting the congregation of people (e.g., church services, athletic events, school, etc.), or prohibition on travel.

Note that quarantine and isolation are similar in some ways, but significantly different in practice. Quarantine is applied both to ill individuals as well as people who have been exposed to a disease, but have not yet begun to show symptoms. Quarantine usually consists of limiting people to their own homes or some other controlled living situation, so that they will not come into contact with unexposed individuals. Isolation, on the other hand, is usually done within hospitals or other medical institutions, and is targeted at people who are clinically sick with the dangerous communicable disease. These patients are maintained in a room that has significant means of limiting the escape of microbes, including the use of negative air pressure to keep contaminated air from the isolation room from escaping. Isolation is expensive, difficult, and is severely limited by the small number of medical isolation rooms found in any given geographic area.

7.6 DISEASE CONTROL MECHANISMS

Understanding the very basics of disease spread is an important first step to comprehending why public health officials will suggest specific interventions. Waterborne or aerosolized droplets are generally the means by which influenza is transmitted. These are small droplets that are expelled during breathing, coughing or sneezing, or some other mechanical means. These also may be called aerosolized or airborne droplets. Microbes can attach themselves to such droplets and these can be directly breathed in, or passed by touching a contaminated surface (that someone coughed on), and then touching a mucous membrane like the mouth or nose.

It is worth noting, in today's global economy, that airplanes are seen as possible conveyors of vectors. We are familiar with animate vectors, e.g., insects and animals, but inanimate transportation devices may serve as conveyors for vectors. For example, it is believed that the arrival of West Nile Virus in the United States was made possible by mosquitoes brought to New York onboard a jet from Egypt (Lanciotti et al., 1999). A few years ago the Severe Acute Respiratory Syndrome (SARS) epidemic was extended from China to the Toronto area of Canada by means of air transportation (Mangili and Gendreau, 2005).

Disease control virtually always starts with an epidemiologic investigation, so that health personnel will know with what they are working. Once the organism and its etiology are known, public health specialists will invoke a control campaign, using one or more of the following tools: vaccination, social distancing (including quarantine and isolation), vector reduction, treatment of infected individuals, and education of the public as to risks and means of preventing transmission.

BOX 7.4 HOW DISEASES SPREAD

- Waterborne droplet (e.g., respiratory diseases)
- Fecal–oral or hand-to-mouth (e.g., diarrheal diseases)
- Water- and food-borne
- Sexual contact (e.g., STDs and HIV)
- Fomites (microbes on inanimate surfaces)
- Blood exposures (e.g., HIV, hepatitis C)
- Vectors (insects, rats, airplanes, etc.)
- Ignorance and bad choices

These are the most common mechanisms employed to control disease spread in a human population. The concepts are simple; implementation is often not. Epidemiologic investigation can be a sizeable undertaking, requiring significant logistical assistance. Quarantine and isolation are legally supported in some jurisdictions, but may not be in others. Even when supported legally, it is difficult to enforce a quarantine. It is best to get the population to cooperate voluntarily.

In summary, the goal of public health, especially with infectious diseases, is to prevent their spread and lessen their impact. This requires the coordinated efforts of most government agencies, the healthcare system, and the population at large. The key is to *prevent spread*. Often the simplest way is as simple as washing one's hands.

7.7 CATASTROPHES AND PUBLIC HEALTH

It doesn't matter what kind of event provokes the catastrophe, all catastrophes lead to significant health impacts (Baxter, 2002). Because the health care system in the United States is consistently working at close to 100% of capacity during normal conditions, it does not take much of an increase in healthcare demand to overwhelm it. Catastrophes provoke a significant increase in demand for healthcare services, while at the same time many of those that provide such services and the facilities in which they deliver this care are themselves affected directly by the event and unable to provide even the normal level of preventive and curative care services. This imbalance can constitute a major challenge for the public and, thus, also for public health and emergency managers, one that will be made infinitely more complicated if emergency managers and public health personnel do not understand each other's skills and needs (Bissell, 2007).

The health effects from a catastrophe can be termed primary or secondary. Primary health effects are those that are caused directly by the event's causal agents (e.g., winds or flooding in a hurricane), or as a result of the direct effects of the event (e.g., the shaking in an earthquake causes buildings and objects within buildings to fall, causing injury. Note that some event types can cause multiple health effects. For example, hurricanes can lead to trauma injury from flying debris, injury from falling structures or trees, and drowning. For more information, see Noji (1997).

BOX 7.5 SAMPLE PRIMARY HEALTH IMPACTS FROM NATURAL HAZARD CATASTROPHES

- Hurricanes: injuries from falling and flying debris, drowning
- Earthquakes: injuries from collapsing structures and landslides
- Tsunamis: drowning, traumatic injury
- Droughts: starvation, thirst

Secondary effects are those that are indirectly caused by the catastrophic event. For example, hurricanes may not directly carry microbial infection, but they can promote disease transmission if humans densely crowd together to seek shelter, or if fresh water supplies are contaminated and ingested without filtration or boiling. Another example would be in chemical or biological agent accidental or intentional release. In this case, there are secondary effects that not only affect victims but also affect public health responders, such as decontamination, contamination, hospital lockdown, positive pressure equipped isolation wards in hospitals, working in personal protective equipment, etc., not to mention having to deal with the worried relatives. In some catastrophes, the human response to the event may provoke more damage to health than the event itself. For example, we now know that placing survivors in tightly packed "refugee" camps, while convenient for care providers and response teams, can be a recipe for significant increases in infectious disease and also can lead to conflict among survivors (Noji, 1997). Let's not forget that there are often secondary mental health issue effects among traumatized populations that can create a demand for mental health services that exceeds the available capacity to deliver these services.

Mass relocation can be another secondary cause of public health deterioration following disasters, and this is one that is particularly likely to be the case in catastrophes. We are now seeing that sea level rise is causing slow onset catastrophes in several low lying countries that are now contemplating having to evacuate the entire populations of these islands to an alternative permanent resettlement location. As an example, research news coverage of the cases of the Maldives and Tuvalu islands in current news sources. (To start your research, see Disappearing Island Nations at: http://itotd.com/articles/499/disappearing-island-nations.)

As will be discussed further in Chapter 8: Mass Evacuation and Relocation, when large numbers of people relocate from one environment to another, they encounter microbes to which their immune-response systems are not accustomed, often resulting in high infection rates. Likewise, the migrants can bring with them diseases that can prove dangerous or fatal to the people already occupying the land into which the migrants are relocating. Further complicating matters, those being forced to evacuate are often under considerable stress, weakening their immune systems and making them more susceptible to disease. For a good historical account of this process, see McNeill (1976) and Diamond (1999).

The importance of the relationship between mass relocation, public health, and catastrophes is that many kinds of catastrophes are likely to result in mass relocations, a problem not normally seen with disasters. This is why some argue that Hurricane Katrina rose to the level, in fact, of a catastrophe, though many would still argue otherwise, based on the number of casualties and the limited impact on the regional and national infrastructure and economy.

There is a long list of secondary causes of health status decreases following catastrophes. The factors in Box 7.6 are only a few of the more prominent causes, and we need to remember that they often combine. For example, loss of employment can lead to loss of housing and sanitation, and decrease in nutritional status. All of these can have important effects on the survivors' emotional status that, in turn, can affect physical health status, and the energy available to work on finding solutions to disaster losses. Because of our lack of national experience with catastrophes, we do not have direct experience with the group psychology/depression of a community of survivors who, because of the size of the event, know that little help will be forthcoming any time soon. It could be that catastrophes will provoke more severe emotional response than disasters because of the extended period of time needed before a significant level of outside help is realized, and because so much of the local community will be dysfunctional ... leading more easily to group depression.

BOX 7.6 DELETERIOUS HEALTH EFFECTS ARE ACCENTUATED IN A CATASTROPHE BY:

- Loss of housing and sanitation
- Loss of food and potable water
- Loss of healthcare facilities and personnel
- Loss of employment
- Loss of organized government support services

If one looks at it from a basic health maintenance perspective, this discussion focuses on the basic components of health in a population. Catastrophes have the capability of taking away, or making scarce, all of these basic components. Once this is realized, one should come to see that all catastrophes are also potential or real public health catastrophes; hence, the importance of involving public health in every stage of catastrophe planning and response.

Knowledge of the determinants of health outcomes is very useful in helping predict what kinds of assistance will be needed and for how long. This is not a calculation that emergency managers will want to take on, but it is useful for them to recognize that the health outcome of a population can be affected by more than just the ferocity of the event, and that some of the determinants can be manipulated to create a better outcome. Likewise, emergency managers should recognize that preevent deficits in nutritional status, education, or vaccine history could lead to considerably worse outcomes and the need for more intervention.

Some determinants of health outcomes following a catastrophe include magnitude and extent of the incident, preevent health status (including nutritional status and immunological experience/vaccines), educational level, and preparedness and training of the public and healthcare workers (Bissell et al., 2004).

This points out once again the multidisciplinary character of the relationship between a population's catastrophe experience and its health outcome, emphasizing the need for a multiagency collaborative, coordinated approach to catastrophe response, as described by McEntire, Maguire, Wachtendorf, and Bissell in this book. See also works by Lagadec (2007) and Kirch et al. (2005) describing the European planning efforts to minimize the health impact of catastrophes, based on a strong emphasis on a broad collaborative approach rather than a singular command and control organizational methodology. See Bissell et al. (2004) for a description of evidence that public health preparedness contributes to a better health outcome in major disasters, including catastrophes.

7.8 PUBLIC HEALTH PRIORITIES

BOX 7.7 PUBLIC HEALTH PRIORITIES IN A CATASTROPHE

- Clean water and sanitation
- Safe and adequate food
- Shelter
- Epidemiologic surveillance/information
- Access to labs
- Access to pharmaceuticals
- Clinical personnel and facilities

The public health priorities listed above can change a bit in terms of which has top priority, depending on the conditions, but all seven entries here are items emergency managers should assume that public health personnel will be asking for. Obviously, some will be considerably more difficult to obtain than others in any given set of circumstances. Note that public health personnel have direct control over very few of these functions, and no control at all over some of the basics (food, housing, and the provision of water). The National Response Framework (NRF) foresees different organizations being responsible for the provision of these basics at the local level, requiring multiagency coordination in order to make available basic necessities that are among the top priorities for public health. Emergency managers will need to play a central role in coordinating between the agencies that carry discrete responsibilities and, where necessary, finding replacements when agencies expected to be available in disasters are not available in a catastrophe.

7.9 INFRASTRUCTURE AND SUPPORT NEEDED FOR PUBLIC HEALTH

The list in Box 7.7 is what public health officials will be requiring or asking for, all of which should be anticipated by emergency managers in their catastrophic planning process. Catastrophes will likely require water supply solutions that are more sustainable than that which might be jerry-rigged in a disaster due to the fact that, in a catastrophe, the "temporary" solution is likely to have to last a lot longer before permanent reconstruction can take place. The same concept holds true for shelter; it may take considerable time before permanent shelter again becomes available, therefore requiring that, where possible, temporary shelter should be replaced with more sustainable transitional shelter as soon as possible. Experience in catastrophe response in developing countries indicates that people will work on repairing their own shelters, if possible, if tools and materials are made available. Whether Americans, especially urban dwellers will, in fact, tend to their own shelter needs, if given the tools to do so, remains unclear.

Note that many items needed to support public health are quite resource-intensive and require considerable expertise in logistics management. Such matters as managing perishable stockpiles of medicines and sterile equipment, and the need for climate-controlled storage, must be effectively addressed. Public health departments may be capable of coordinating the influx of qualified medical personnel into the catastrophe zone, but will likely need help with all of the logistics, transport, and communications functions listed in Box 7.8. The U.S. Department of Health and Human Services (HHS) maintains medical and public health response teams, some of which consist of volunteers within the National Disaster Medical System (NDMS), and some of whom are full-time uniformed personnel of HHS who are trained and maintain readiness status. Public health responders and investigators based around the country have been trained at the CDC.

BOX 7.8 INFRASTRUCTURE AND SUPPORT NEEDED FOR PUBLIC HEALTH

- Trained epidemiologic field personnel with assistants, computers, means of communication, transportation, mobile lab equipment, shelter
- Means of conveying specimen samples to a qualified reference lab for diagnosis
- Pharmaceutical supply and distribution network
- Temporary or permanent hospitals, clinics, diagnostic and treatment equipment, all with power, water, sanitation, and food
- Clinicians of all levels, particularly those with primary care skills, and ancillary support staff

7.9.1 The National Disaster Medical System (NDMS)

The NDMS is a federally coordinated system that augments the nation's medical response capability (Figure 7.1). The overall purpose of the NDMS is to supplement an integrated national medical response capability for assisting state and local authorities in dealing with the medical

FIGURE 7.1 The National Disaster Medical System (NDMS) seal.

impacts of major peacetime disasters and to provide support to the military and the Department of Veterans Affairs medical systems in caring for casualties evacuated back to the United States from overseas conventional armed conflicts.

The National Response Framework utilizes the NDMS, as part of the HHS, Office of Preparedness and Response, under Emergency Support Function (ESF) #8: Health and Medical Services, to support federal agencies in the management and coordination of the federal medical response to major emergencies and federally declared disasters including:

- Natural disasters
- Major transportation accidents
- Technological disasters
- Acts of terrorism including weapons of mass destruction events

NDMS's mission is to temporarily supplement federal, state, and local capabilities by funding, organizing, training, equipping, deploying, and sustaining a specialized and focused range of public health and medical capabilities. Components of NDMS include:

- Medical response to a disaster area in the form of personnel, teams, and individuals, supplies, and equipment.
- Patient movement from a disaster site to unaffected areas of the nation.
- Definitive medical care at participating hospitals in unaffected areas.

These federal personnel constitute a significant potential public health workforce in a catastrophe, but they are limited by two factors: (1) they often lack local knowledge and contacts, and (2) they have only short-term self-sufficiency in terms of transportation and supplies, after which they may depend on local coordinators to make it possible for them to continue functioning. Given the complexity and magnitude of catastrophes, such teams will only be effective if they benefit from multiagency, multidisciplinary coordination and support.

7.10 ROLE OF SURGE CAPACITY PLANNING IN CATASTROPHES

Surge capacity is a measure of the ability of a hospital or clinical care system to rapidly adjust to caring for a sudden increase in the number of patients. This has become a primary concept in public health agencies as they conduct pandemic flu planning. Other kinds of disasters also may cause surges in patient numbers and, if they coincide with a decrease in the numbers of facilities available (due to damage, destruction, loss of supporting infrastructure, or loss of personnel), the concept of rapidly increasing clinical care capability in surviving facilities might be viable, at least in disasters with limited geographic spread. In catastrophes, the surge capacity would have to be seen as applying to the national level. In a pandemic, everyone would be overwhelmed and the amount of surge capacity available is miniscule compared to the need that might well exist. The NDMS, CDC, as well as other (HHS) and other teams mentioned previously all have plans for bringing health personnel into affected areas and, in the case of NDMS, transporting patients out of the affected areas to receive care in other parts of the country. Whether these resources and plans would be sufficient to meet the need depends on the circumstances of the catastrophe, but none of them will be effective in a catastrophe without substantial assistance from, and coordination with nonpublic health agencies and organizations, including the military, that support and complement their efforts. So, while we talk about *surge capacity* at the national level for catastrophes, it all still depends on coordination at the local level to make the most effective possible use of scarce resources. (For more information on NDMS, go to: http://www.hhs.gov/aspr/opeo/ndms/index.html.)

There are two basic strategies or mechanisms for creating a surge of capacity. The first is to enhance the number of clinicians and other workers taking care of patients or victims by using staff who are already internal to the organization (e.g., hospital, emergency management agency, fire service), by bringing in all available staffing shifts at once. This strategy might be called an *internal surge*. A second strategy is to bring in outsiders and incorporate them into the organization's tasks. This might be termed an *external surge* strategy and a good example of it might be the disaster operations of the American Red Cross, in which the organization plans for and trains for the utilization of outside volunteers in disaster responses. Most surge capacity planning in the healthcare sector assumes an internal surge strategy or, at best, very limited use of outsiders. State licensing and credentialing requirements are often cited as a limiting factor in bringing in external healthcare workers. The severe conditions of catastrophes may well benefit more from a preplanned external surge capacity strategy, given the extreme levels of need at the same moment that internal resources are highly limited. Emergency managers can help move jurisdictional health authorities toward focusing on external strategies during catastrophic planning by providing planning assistance and emphasizing the potential impact a catastrophe may have on the community's ability to meet healthcare needs of its population.

7.11 EMERGENCY MANAGEMENT: PUBLIC HEALTH COLLABORATION IN CATASTROPHES

Public health authorities have scientific capabilities and, in some places, statutory responsibility, but virtually no logistics, transport, communications, or law enforcement resources with which to conduct their work in a catastrophic working environment. In addition, public health authorities have experience coordinating within the health sector, but not with outside agencies. From this, we can deduce that public health authorities will need to depend on emergency management agencies in order to be as effective as possible in catastrophes. This normally occurs within ESF #8, Health and Medical Services under the NRF. The responsibilities of public health will be broader than the resources public health authorities can command. Emergency management agencies will need to provide significant assistance in logistics, resource management, and coordination with outside groups in order to have the greatest positive impact on meeting the public's health needs following a catastrophe. The reader should be thinking of ways in which the similarities between public health and emergency management can help facilitate coordination that is needed between the two disciplines, as long as both public health and emergency management recognize and overcome some of the differences in vocabulary, nomenclature, and decision-making styles.

Here, we recognize that emergency management agencies also need the assistance of public health entities to conduct needs assessments and help with priority planning and implementation. We also remember the important role that public health plays in assuring the well-being of emergency managers and emergency response personnel, a role that will become all the more important in catastrophes with drawn-out periods of operations.

7.12 DISCUSSION QUESTIONS

1. Try to learn about how your community plans on handling a future public health emergency, possibly a pandemic influenza outbreak. What legal authorities do public health, emergency management, and law enforcement personnel have to manage such an emergency? Has there ever been a public health emergency drill or exercise within your community?
2. In an epidemic scenario, why is it not sufficient to just treat everybody who comes down with the disease?
3. Assume a major earthquake along the New Madrid Seismic Zone (NMSZ). There are more than 100,000 injured. Hospitals and clinics are down, water and sewer systems are severed, road and bridge failures block food and medicine deliveries. What health concerns emerge from this scenario and how must emergency management and public health collaborate?

REFERENCES

Baxter, P.J. 2002. Catastrophes: Natural and manmade disasters. In *Conflict and catastrophe medicine,* eds. J. Ryan P.F. Mahoney, I. Greaves, and G. Bowyer, Chap. 2. London: Springer Verlag.

Bissell, R.A. 2007. Public health and medicine in emergency management. In *Disciplines, disasters and emergency management*, ed. D.A. McEntire, Chap. 16. Springfield, IL:, Charles C. Thomas Publisher.

Bissell, R.A., L. Pinet, M. Nelson, and M. Levy. 2004. Evidence of the effectiveness of health sector preparedness in disaster response: The example of four earthquakes. *Family and Community Health* (special issue: *Disaster Management in Public Health*)27(3): 193–203.

Diamond, J. 1999. *Guns, germs and steel: The fates of human societies.* New York: Norton Books.

Friis, R.H. and T.A. Sellers. 2004. *Epidemiology for public health practice.* Boston: Jones and Bartlett Publishers.

Kirch, W., B. Menne, and R. Bertolini. 2005. *Extreme weather events and public health responses.* London: Springer Verlag.

Lagadec, E. 2007. *Unconventional crises, unconventional responses: Reforming leadership in the age of catastrophic crises and hypercomplexity*. Baltimore, MD: Johns Hopkins University, Center for Transatlantic Relations.

Lanciotti, R.S., J.T. Roehrig, V. Deubel, J. Smith, M. Parker, K. Steele, B. Crise, K.E. Volpe, M.B. Crabtree, J.H. Scherret, R.A. Hall, J.S. MacKenzie, C.B. Cropp, B. Panigrahy, E. Ostlund, B. Schmitt, M. Malkinson, C. Banet, J. Weissman, N. Komar, H.M. Savage, W. Stone, T. McNamara, and D.J. Gubler. 1999. Origin of the West Nile virus responsible for an outbreak of encephalitis in the northeastern United States. *Science* Dec. 17, 286(5448): 2333–2337.

Landesman, L.Y. *Public health management of disasters*. 2001. Washington, D.C.: American Public Health Association.

Mangili, A., and M.A. Gendreau. 2005. Transmission of infectious diseases during commercial air travel. *Lancet* 365(9463): 989–996.

McNeill, W.H. 1976. *Plagues and peoples*. New York: Bantam Doubleday Dell Publishing Group.

Noji, E. (ed.) 1997. The nature of disaster: General characteristics and public health effects. In *The public health consequences of disaster*. Oxford, U.K.: Oxford University Press.

Sundnes, K.O., and M.L. Birnbaum. 2003. Health disaster management: guidelines for evaluation and research in the Utstein style. *Prehospital and Disaster Medicine* 17(Suppl. 3).

CHAPTER 8

Mass Evacuation and Relocation

8.1 LEARNING OBJECTIVES

As a result of this chapter, you should be able to:

- Contrast between preevent and postevent evacuation, and relocation.
- Analyze major issues that may arise in the event of mass evacuation and resettlement, including possible methods of addressing these issues.
- Analyze the relocation continuum (e.g., the different things/events that occur during relocation, such as emergency sheltering, temporary housing, and long-term housing and relocation postcatastrophe).
- Appraise transportation modes for both evacuation and relocation.

8.2 KEY TERMS AND PHRASES

- Continua of Displacement
- Demographic movement
- Displacement
- Distributive justice
- Ecosystem services
- Environmental refugees
- Escape
- Flight
- The four stage framework for resettlement
- Framework convention on climate change
- Human rights
- Impoverishment Risks and Reconstruction Model

- Involuntary displacement and resettlement
- Forced migration
- Mass displacement
- Mass evacuation
- Mass migration
- Mass relocation
- Mitigation
- Procedural justice
- Refugee
- Resettlement
- Resettlement Action Plans (RAP)
- Social vulnerability
- Sustainable livelihoods
- Temporary mass shelter
- United Nations Framework Convention on Climate Change (UNFCCC)
- United Nations Guiding Principles on Internal Displacement

8.3 INTRODUCTION

The purpose of this chapter is to present an overview of the field of displacement and resettlement research, focusing on the development of conceptual approaches, policy positions, and practice problems in the various forms of displacement and resettlement associated with catastrophic events. The range of forms that catastrophic-forced displacement and resettlement are projected to take is considered. Emphasis is placed on developing an understanding of the factors that generate both the short- and long-term risks and consequences in major dislocations, deriving understanding from data, and perspectives from other forms of displacement and resettlement, including conflict and development-caused relocations. This chapter also identifies and analyzes the key components of resettlement planning as developed for infrastructural projects, such as large dams, assessing their utility for crafting appropriate standards and strategies for potential future mass relocation.

Mass relocation after a catastrophe, involving the physical displacement and resettlement of people, is an extremely complex process, which if not properly planned and managed (with the full participation of affected people) may result in unneeded additional long-term hardship for the displaced as well as potential conflict with resident populations and environmental damage in locations in which they are resettled. This discussion starts with an overview of the field of displacement and resettlement research, focusing on the development of theoretical approaches, policy initiatives, and practice problems in the various forms of displacement and resettlement. The range of forms that environmentally forced displacement and resettlement have taken and are projected to take also are considered.

Over the past half-century, researchers on development-induced displacement, refugee studies, and disaster research (Hansen and Oliver-Smith, 1982; Cernea, 1996; Cernea and Kanbar, 2002; Turton, 2003) have learned that involuntarily displaced peoples face many similar challenges. Although the places and peoples are geographically and culturally distant and the socio-political environments and causes of dislocation dissimilar, there emerge a number of common concerns and processes. Displaced people of all descriptions must cope with the consequent stresses and the need to adapt to new or radically changed environments. All may experience hardship, loss of homes, jobs, and the breakup of families and communities. All may suffer the endangerment of structures of meaning and identity. All must mobilize social and cultural resources in their efforts to reestablish viable social groups and communities and to restore adequate levels of material and cultural life. If done right, those that are displaced regain a sense of identity.

The discussion, therefore, identifies the key components of resettlement planning as developed across a number of fields, assessing the utility of various approaches for crafting appropriate standards and strategies for future catastrophe-driven mass relocation. Central to these tasks are the issues of rights, poverty, vulnerability, and other forms of social marginality that are intrinsically linked to displacement. An important additional notion is that the displaced must be seen as active social agents with their own views on rights and entitlements, which have to be considered in any displacement and in the planning and implementation of resettlement projects.

8.4 DEFINING MASS RELOCATION

BOX 8.1 THE TWO PROCESSES OF MASS RELOCATION

- **Displacement:** Due to the occurrence (or imminent occurrence) of a disaster, including climate-induced environmental change, conflict, or development, people are forced to leave their place of abode because it has been rendered uninhabitable either temporarily or permanently.
- **Resettlement:** The reestablishment of displaced peoples in a new location with appropriate settlement design, housing, services, and an economic base to enable the community to reconstitute itself and achieve adequate levels of resilience to normal social, economic, political, and environmental variation.

Mass relocation, also frequently referred to as *forced migration* or *involuntary displacement and resettlement*, refers to the uprooting of large numbers of people from their home locations. Although the term *mass* refers to large numbers, it is vague and not well defined in its application. Nonetheless, we should not fail to recognize that future potential displacements of enormous size are projected for the not too distant future. Even today, the several hundred thousand people remain dispersed from New Orleans and surrounding parishes by the combination of Hurricane Katrina, and an ad hoc and poorly conceived government response constitutes a mass displacement.

However different the driving forces and policies may have been for forcibly uprooted people, recovery and reconstruction take place in a new setting, generally far from familiar environments and people. In other words, getting to where they are going does not solve the problem. They may have stopped moving, but that is just the beginning of another process—resettlement. In all too many cases, resettlement, particularly when done at the community level with support for high levels of government, ends up becoming a secondary crisis.

Therefore, when catastrophes, conflicts, or development damage or destroy communities, uprooting people, displacing them far from homes and jobs, the process of recovery is made doubly complex. Some mass relocations will involve sudden rapid onset events that evoke, at initial stages, elements of emergency management strategies, such as mass evacuation and temporary mass shelters. Other approaches to deal with mass relocations may resemble the resettlement of political refugees in strategies to integrate the displaced into existing communities.

Still other forms will be the result of planned mitigation projects, such as making way for the inundation area behind a new dam, and will draw on models from development-forced resettlement, community development, and urban planning. Some mass relocations may involve several of these forms of displacement and resettlement. Finally, some mass relocations will constitute simply mass migrations, evoking very little formal institutional response. The topic of this discussion, thus, requires input from all of the phases of emergency management, and many social, scientific, and management disciplines.

In some circumstances, because catastrophes involve different time/space scales (lasting longer, encompassing wider areas), crossing ecological, jurisdictional, and national boundaries, and impacting heterogeneous populations, they will require multiple strategies and often inter- and multinational efforts and cooperation. At the same time, mass relocation may involve masses of people, but responses will need to address culturally and socially defined constituent population groups. Regardless, uprooted people generally face the daunting task of rebuilding not only personal lives, but also those relationships, networks, and structures that support people as individuals that we understand as communities. The social destruction wrought by these phenomena takes place at both the individual level and at the community level. In most cases, solutions must be durable or there is often little hope of return to their homeland.

Given the lack of research on catastrophes and mass relocations, particularly in contemporary times, much of what follows is drawn from research and practice in the fields of refugee studies, disaster research, migration, planning, and development-forced displacement and resettlement.

8.5 COMPLEXITY AND CAUSATION

Since the 1980s, researchers have linked the issue of catastrophic environmental change with human migration, explicitly designating as *environmental refugees* people who are forced to leave their homes, temporarily or permanently, due to the threat, impact, or effects of a hazard or environmental change (El-Hinnawi, 1985). Other scholars attribute the displacement of people to a more complex pattern of factors including political, social, economic as well as environmental forces (Wood, 2001; Black, 2001; Castles, 2002). Catastrophes resulting from natural hazards are seen to cause temporary displacement. Indeed, if permanent migration does occur as the result of a catastrophe caused by a natural hazard, it is seen as more the result of deficient responses of weak or corrupt states rather than an altered environment as expressed in the form of a natural hazard impact. Certainly, Hurricane Katrina exemplifies this perspective. Black's (2001) critique that focusing on environmental factors as causes of migration often obscures the role of political and economic factors is well-taken, and echoes the position held by most disaster researchers today that focusing solely on agents that produce the upheaval reveals little about the political or economic forces that together with agents produce catastrophes or, for that matter, any forced migration that might ensue.

Seeking single agent causality is always highly problematic. There are two fundamental questions regarding causality. The first asks what empirical evidence is required for legitimate inference of cause–effect relationships. The second suggests that if we are willing to accept causal information about a phenomenon, what kinds of inferences can be drawn from that information (Pearl, 2000)? The key word here is "inferences." Clear and direct relationships of causality are hard to come by. In the strictest sense of the word, if *A* causes *B*, then *A* must always be followed by *B*. In common parlance, when we say *A* causes *B*, as in smoking (*A*) causes cancer (*B*), what we should really say is that smoking causes an increase in the probability of cancer (Spirtes et al., 2000).

In other words, in the case of catastrophes, *A* increases the risk of *B*, or forced migration.

Therefore, it is difficult to point to the environment, even in catastrophes, as the single cause of anything. By the same token, eliminating environment factors as the single cause of forced migration hardly warrants discounting them as one of a multiplicity of forces at work in generating mass relocation. It is important to remember here that a catastrophe also is not defined in terms of its event aspect only, but in terms of both the processes that set it in motion and the postevent processes of adaptation and adjustment in recovery and reconstruction. Forced migration can be part of the process prior to the event or after, but it is not inevitable. We know that catastrophes are not caused by a single agent, but by the complex interaction of both environmental and social features and forces.

By the same token, outcomes of catastrophes are rarely the result of a single agent (i.e., a hurricane), but are brought about by multiple complex and intersecting forces acting together in a specific social context that is complex in its own right. Seeking single causes for a complex outcome is usually difficult in any context and particularly so with forced migration, whether the obvious "cause" is international or civil conflict, development projects, or natural or technological disasters.

8.6 UNDERSTANDING MASS RELOCATION

BOX 8.2 DISPLACEMENT AND RESETTLEMENT VOCABULARY

- **Conflict:** camps, international and individual, family focused
- **Disaster**-induced displacement: Evacuation, temporary shelters, reconstruction
- **Development**-induced displacement and resettlement: Constructing new settlements

The social scientific literature on displacement and resettlement is clustered around three themes: civil and military conflicts, disasters (by extension, catastrophes), and development projects. The relatively scant literature from disaster/catastrophe-driven displacement focuses largely on temporary shelters, with a few cases dealing with permanent resettlement of small communities (Oliver-Smith, 1991; Perry and Mushkatel, 1984). The research, related to conflict-driven uprooting, focuses largely on temporary camps, repatriation and individual/family refugee resettlement to foreign countries (Martin et al., 2005). The literature on development-forced displacement and resettlement generally deals with resettlement of communities of varying size and, sometimes, whole regions are affected by large-scale infrastructure projects (Cernea, 1996; Scudder and Colson, 1982; de Wet, 2006). It is clear that catastrophe-driven mass relocation must draw on these other fields for insights into how best to understand and respond to the potentially large-scale displacements projected for the not too distant future. This research is also being complemented by a growing concern regarding Internally Displaced Persons (IDP) (Deng and Cohen, 1999; Koser, 2007). Unlike refugees who cross national borders and benefit from an established system of international protection and assistance, those forcibly uprooted within their own countries, known as IDPs, lack predictable structures of support. There are currently 25 million IDPs worldwide in at least 50 countries (Deng and Cohen, 1999).

For example, in the United States, government to nongovernmental organizations (NGOs) are largely entrusted with refugee resettlement. There is little attention paid to the idea of community of origin. There is, however, a debate between NGOs that favor resettling refugees with co-ethnics so that they can help each other with language and employment issues and those that believe in dispersing the displaced so that they will assimilate, including learning English faster (Hansen, 2005). The lack of attention to issues of community may stem from the position that Americans have become ideologically distanced from the idea of community as something people need. Americans have been portrayed as seeing themselves as eminently mobile, able to adapt easily to new homes, new jobs, and new networks. The degree to which that contention is true for Americans may be debated, but it is certainly not the case for many of the world's people. However, the discourse of displacement and resettlement in American society, that is, the choice of terminology and the scale or unit of analysis most frequently addressed, is at the level of individuals and families, whereas most large-scale displacement very frequently involves communities. This is significant, particularly with regard to losses, because what often becomes lost is the community network that enabled people to access resources; not just material resources, but social and emotional support that in stressful times in the displacement of communities becomes all the more a significant issue. The community is more than the sum of the total number of individuals and the loss of community for displaced people, particularly when the loss is the outcome of aid policies that do not take community into account, can be devastating.

When an entire community is resettled, it is not simply lifted up and set down whole in a new site. In most cases the community is reconfigured in specific ways. Most resettlement projects, particularly in the developing world, directly or indirectly further two fundamental processes: (1) the expansion of the government's role and (2) integration into regional and national market systems. Neither of these processes of inclusion is particularly simple or straightforward, but in most cases, they produce a restructuring of social, economic, and political relationships toward the priorities of the larger society. In many respects, resettlement will not necessarily destroy "local cultures" as much as it appropriates them and restructures them in terms of values and goals often originating from far beyond the local context. Such a process involves the reduction of local culture, society, and economy from all their varied expressions to a narrow set of institutions and activities that make them compatible with the purposes of the larger society (Canclini, 1993).

If we are to both understand and respond effectively to potential mass displacements from global climate change, we need to identify those pertinent sources of theory and information that can inform appropriate policy formation and practice. The process of resettlement in cases of involuntary uprooting has proved to be a particularly challenging one. In point of fact, the record of successful resettlement projects is generally considered to be far from being fully successful. The vast majority of these projects, whether from disasters/catastrophes, development, or conflict-driven displacement, have left local populations permanently displaced, disempowered, and destitute. For the vast majority of the displaced, the causes of dislocation and the uprooting process itself are nothing less than catastrophic both at the personal level and at the level of community. These forces, natural and technological disasters, political conflicts, and large-scale development projects, are what Oliver-Smith (2006) calls "totalizing phenomena" in their capacity to affect virtually every domain of human life.

8.7 HISTORIC PERSPECTIVE OF CATASTROPHES AND MASS RELOCATION

A frequent response to catastrophe throughout history has been mass displacement, but most cases have not evoked significant or effective policy or practice responses. In general, we must look to history for detailed discussions of these processes. There are relatively few contemporary studies of catastrophe-caused community displacement and resettlement that have been the result of planned policy and action in the United States and elsewhere, but these have been relatively small events involving smaller populations. Their utility in understanding and responding to mass relocations remains to be assessed.

8.7.1 Irish Potato Famine

In the Irish potato famine, formal institutional responses actually exacerbated conditions, forcing huge numbers of starving peasants to leave the country by whatever means was available, often to North America or Australia as indentured servants (Figure 8.1). A significant percentage of them died while in transit. There were few formal institutional measures undertaken to assist them in either the displacement or the resettlement processes and the hardships and discrimination they endured in their new locations are well documented.

FIGURE 8.1 Antique engraving of "Emigrants leaving Ireland." (From *Illustrated History of Ireland*, 1868.)

8.7.2 Great Mississippi Flood of 1927

The Great Flood of 1927 in the lower Mississippi Valley (Figure 8.2) displaced nearly 700,000 people, approximately 330,000 of whom were African Americans who were subsequently interned in 154 relief "concentration camps" where they were forced to work. Although there were many reasons for African Americans to leave the South, the flood and its consequences, especially the forced labor in the camps, were the final motivation for migrating for thousands (Barry, 1997, p. 417) (Figure 8.3).

8.7.3 The Dust Bowl

During the great economic depression of the 1930s, several years of inadequate rainfall and elevated temperatures in the Great Plains resulted in the widespread failure of small farms. This event was known as the *dust bowl* and it produced a migration of close to 300,000 people from the region to the west coast of the United States. These migrants were often called "Okies" as many came from Oklahoma. Those that relocated to

FIGURE 8.2 Downtown Greenville, Mississippi, on April 30, 1927, six days after the levee break. (Photo courtesy of the Mississippi Department of Archives and History.)

FIGURE 8.3 Flood refugees evacuated to "tent city" on the hills of Vicksburg, Mississippi. (Photo courtesy of the Mississippi Department of Archives and History.)

FIGURE 8.4 California has not been kind to this family. Their two-year-old child died of exposure during the winter. They asked relief authorities to help them return to Oklahoma. (U.S. Library of Congress.)

the west coast attempted to assimilate into the local agrarian-based society with varying levels of success. Some grew so disenchanted that they eventually returned to the home state, preferring to face the challenges of repeated failure of crops (Figure 8.4).

In response to this crisis, the federal government created the Farm Security Administration (FSA) that focused on assisting farmers and displaced farmers (Figure 8.5). Initially created as the Resettlement Administration in 1935 as part of President Roosevelt's New Deal, FSA was an effort during the Depression to combat rural poverty in the United States. The FSA stressed "rural rehabilitation" efforts to improve the lifestyle of sharecroppers, tenants, and very poor landowning farmers, and a program to purchase submarginal land owned by poor farmers and to resettle them on group farms on land more suitable for efficient farming. Critics, including the Farm Bureau strongly opposed the FSA as an experiment in collectivizing agriculture, i.e., in bringing farmers together to work on large

FIGURE 8.5 The Department of Agriculture Farm Security Administration logo.

government-owned farms using modern techniques under the supervision of experts. The program failed because the farmers wanted land ownership; after the Conservative Coalition took control of Congress in 1937, it transformed the FSA into a program to help poor farmers buy land, and it continues today as the U.S. Department of Agriculture's Farmers Home Administration.

One of the more tangible activities of the FSA was the construction of resettlement "camps." These camps tended to offer a communal living setting, where residents had to share common social halls and sanitary facilities. Note the large meeting hall in the center of Figure 8.6. Another benefit of these camps was that they were considered to be federal reservations and local law enforcement, which often was accused of harassing the camp occupants, was barred from entering the camps.

If you are interested in learning more about the Dust Bowl, there are many online resources. One of the more interesting Web sites is: http://www.livinghistoryfarm.org/farminginthe30s/water_02.html, which contains a number of streaming videos of interviews with people that lived through the Dust Bowl years. (Quicktime® streaming video software is needed to watch the videos.)

FIGURE 8.6 Highly touted by the New Dealers, the Farm Security Administration camps were as much a political symbol as a practical answer to the problems of farm labor. Intended to foster cooperative values as well as provide emergency shelter, the design called for platform camp sites surrounding a common washroom and recreational center. (U.S. Library of Congress.)

8.7.4 Hurricane Katrina

Hurricane Katrina in 2005 uprooted about 1.5 million people, 300,000 of whom are expected to remain permanently displaced (Figure 8.7). Their displacement, however, was not due to environmental reasons alone, but to inappropriate policy, incompetent practice, and the political economy of reconstruction as well. Some of the displaced took advantage of economic opportunities in the communities to which they were relocated and see no economic justification to return to New Orleans, though they continue to suffer from the effects of social dislocation.

FIGURE 8.7 Evacuees from Hurricane Katrina wait in lines to board aircraft for evacuation at the New Orleans airport on December 2, 2005. (Photo by Michael Rieger/FEMA.)

8.8 GLOBAL CLIMATE CHANGES AND MASS RELOCATION

Global climate change has been projected to become one of the major driving forces of mass relocations in the near future. The recent reports from the Intergovernmental Panel on Climate Change (IPCC, 2007) affirm that human-induced factors are responsible for generating significant increases in temperatures around the world. Among the consequences of this rise in temperature are supposedly increases in the rate of sea level rise, increases in glacial, permafrost, Arctic and Antarctic ice melt, more rainfall in specific regions of the world and worldwide, more severe droughts in tropical and subtropical zones, increases in heat waves, changing ranges and incidences of diseases, and more intense hurricane and cyclone activity, as described earlier in this book.

Moreover, many of these changes are compounding each other to accelerate the rates at which they are proceeding. All of these changes are projected to affect natural systems globally, inducing alterations in hydrological, terrestrial, biological, and aquatic environmental subsystems. All of these changes also have great potential for generating processes that may lead to the uprooting of large numbers of people, forcing them to migrate as individuals and families or permanently displacing them and/or relocating them as communities. Global climate changes, in addition, are projected to also combine with other factors, such as environmental contamination, to drive people from their homelands. While all of the changes mentioned have the potential to uproot people, there are basically three major expressions of global climate change that will principally contribute to the forces that uproot people: loss of ecosystem services, loss of land, and increased intensity and frequency of climate based natural disasters. Moreover, each of these forces can be said to interact with each other to compound and intensify the effects of all of them.

8.8.1 Loss of Ecosystem Services

First, global climate change is expected to seriously alter the availability and access of ecosystem services. Human life in specific environments is maintained by the provision of a number of ecosystem services, including food, water, fuel, and nutrition, as well as those cultural elements, largely spiritual and/or aesthetic, that sustain communities through expressive links to natural features (Renaud et al., 2007). However, in combination with other factors, such as nutrient pollution, over exploitation, invasive species and diseases, as well as demographic, economic, socio-political, and cultural factors, global climate change may strain the resilience of local socio-ecological systems to the degree that they no longer are capable of providing the necessary eco-system services required for continued habitation by humans. The degradation of local ecosystems compounded by climate changes can bring about loss of access to sufficient resources and other ecosystem services. This also will uproot people and communities, forcing them to migrate in search of the minimal requirements for survival that their home environments can no longer provide. Drylands, estimated to cover approximately 41% of the Earth's land surfaces and home to roughly two billion people, are particularly prone to the process of desertification that leaves those environments so depleted of ecosystem resources and services that they will no longer sustain human life.

8.8.2 Loss of Land

Although the loss of land might be considered to be part of the loss of ecosystem services, we treat it separately because it constitutes not only the loss of resources necessary for life, but also the actual disappearance of terrain. Driven largely by sea level rise, the people who occupy regions in the low elevation coastal zones (between 1 and 30 feet above sea level) of the world are becoming vulnerable to the permanent inundation of their homes and livelihoods. These regions contain 10% of the current world population and 13% of the urban population, including almost two-thirds of cities with populations larger than five million people. Recent research estimates that:

> [I]n all, 634 million people live within such areas—defined as less than 30 feet above sea level—and that number is growing. Of the more than 180 countries with populations in the low-elevation coastal zone, about 70% have urban areas of more than five million people that extend into it, including Tokyo; New York; Mumbai, India; Shanghai, China; Jakarta, Indonesia; and Dhaka, Bangladesh. Indeed, about 75% of all the people residing in low-lying areas are in Asia and the most vulnerable are the poor. (McGranahan et al., 2007).

Over the past five years, communities in coastal areas in Alaska, the South Pacific, and the Gulf Coast of the United States are facing the threat of community-wide displacement and resettlement. The physical and social processes recently triggered by Hurricane Katrina on the Gulf Coast of the United States underscore the threat of this emerging reality. While some displacement is likely to be gradual as coastal land is increasingly inundated over the coming years, elevated sea levels will also increase the impacts of tropical storms, creating sudden, devastating disasters and possibly catastrophes that were not forecast to occur before the sea level rose.

8.8.3 Increasing Intensity and Frequency of Climate Driven Disasters

Hurricane Katrina is also emblematic of the third major driver of environmental displacement and resettlement. With the increase in surface temperature over the last 50 years, the levels of damage from extreme weather events have also increased. On a global scale, losses from natural disasters have increased dramatically over the last half-century, particularly so since the mid-1980s (Munich Reinsurance Co., 1999, p. 16). Global climate change will increase the risk of stronger tropical cyclones (hurricanes, typhoons, and cyclones) with higher storm surges, which when combined with rising sea levels, will extend the onshore impacts of coastal flooding much farther inland, particularly in the low coastal elevation zone. While damage from high winds in tropical cyclones can be devastating, the risk of increased flooding has the highest potential for the displacement and resettlement of communities. With increasing evidence that what used to be 'a once-in-a-100-year event' is becoming more common"(Huq et al., 2007, p. 4), resulting displacements could become permanent as more and more coastal land is lost to the sea or eliminated as habitable zones because of tidal storm surges.

8.8.4 Environmental Change and Forced Migration

Environmental change does not necessarily undermine human security in the absence of poverty, lack of economic opportunity, lack of government support, good governance, and social cohesion with surrounding groups, but at present we know very little about the interplay between environmental change, ecological systems, socio-economic vulnerability, and patterns of forced migration.

Climate change may create a political imperative to bring about socio-economic reforms to address deeply rooted poverty/vulnerability issues. What is clear is that we cannot live sustainably in our environment unless we live in a just and equitable relationship with one another. Climate mitigation and adaptation strategies are just wishful thinking unless long-standing global inequalities/imbalances are addressed in the process.

8.9 DEFINING THE DISPLACED

The debate over the term and category of *environmental refugees*, with claims of millions of environmental refugees being produced versus counterclaims that the evidence is uneven, unconvincing, and counterproductive, has been active since the 1980s. Myers (1997) has asserted that recent human-induced environmental change, such as desertification, deforestation, or soil erosion, compounded by natural and man-made disasters, could force as many as 50 million people to migrate from their homes by 2010. Other researchers dispute the accuracy of the term *environmental refugee,* finding it misleading. They attribute the displacement of people to a complex pattern of factors including political, social, economic, as well as environmental forces (Wood, 2001; Black, 2001; Castles, 2002). Environmental disruptions, including natural disasters/catastrophes, are seen to cause temporary displacement, but not some idea of authentic, i.e., permanent, migration. Indeed, if permanent migration does occur as the result of a catastrophe, it is seen as more the result of deficient responses of weak or corrupt states rather than the environment as expressed in the form of a natural hazard impact.

Black's (2001) critique that focusing on environmental factors as causes of migration may obscure the role of political and economic factors is well taken, and echoes the position held by most disaster researchers today. Focusing solely on agents reveals little about the political or economic forces that together with agents produce catastrophes or, for that matter, any forced migration that might ensue. But these objections turn aside the fact that the environment, and its resources as well as its hazards, is itself socially constructed and is always channeled for people through social, economic, and political factors, even in the best of times (Oliver-Smith, 2002). The environment cannot be separated from society to isolate it as a single cause. Nature and society, where they co-exist, produce socially constructed environments composed of both natural features and social constructions that are mutually complementary. In that sense, the environment cannot be isolated as a single cause in most cases because it is interwoven with society. It is equally important to remember here that a catastrophe is also no longer defined in terms of its event aspect only, but in terms of both the processes that set it in motion and the postevent processes of adaptation and adjustment in recovery and reconstruction.

8.10 HUMAN RIGHTS DIMENSIONS OF MASS RELOCATION

Human rights are a central issue in any mass relocation. In cases of conflict and development, the intentionality and coercion exercised in the displacement is generally accepted as constituting a gross human rights violation. In a catastrophe, the social vulnerability of those most affected is often the result of a compendium of denial of fundamental human rights. In climate change, equity is a core principle of the Framework Convention on Climate Change.

Procedural justice refers to who makes the crucial decisions on climate change. Who decides what strategies to adopt? And, indeed, who is responsible for the decisions that produced actions that resulted in climate change? Indeed, the people least responsible for climate change, people in the developing world, are those who will likely suffer the greatest impacts.

Distributive justice refers to inequities in the distribution of risk, vulnerability and impact that have resulted in selective victimization. Most of the people most vulnerable to climate change are in the developing world. Climate change and our adaptations to it, such as relocation as a mitigation strategy, threaten to exacerbate exactly those forces that cause present insecurities in developing countries and will likely increase that insecurity in the future possibly causing significant social and economic turbulence.

Both our adaptations and the distribution of impacts are distributed unequally in terms of their effects on attempts by nations to reduce poverty and vulnerability. Adaptations to climate change may exacerbate past injustices (underdevelopment, exploitive development, and colonialism) that are, in effect, the conditions that produced the patterns of underdevelopment, exploitive development, and vulnerability to climate change and other disasters (Adger, Paavola, and Huq, 2006).

8.11 MASS RELOCATION AND THE LEGAL STATUS OF THE DISPLACED

Although the issue of environmentally displaced peoples has generated significant debate over the past 20 years, appropriate policies pertaining to environmentally displaced peoples or other IDPs have yet to attain legal status. Moreover, according to the International Federation of Red Cross and Red Crescent Societies (IFRC, 2004):

> ... there are no well recognized and comprehensive legal instruments which identify internationally agreed rules, principles and standards for the protection and assistance of people affected by natural and technological disasters. As a result, many international disaster response operations are subject to ad hoc rules and systems, which vary dramatically from country to country and impede the provision of fast and effective assistance—putting lives and dignity at risk (IFRC, 2004, p. 1).

The category *refugee* with all its attendant rights also still applies only to a very specifically defined group of people who, in fleeing for their lives, have crossed an international border. However, over the past decade there has been increasing concern regarding IDPs and their rights, and there is increasing recognition that the causes of displacement and resettlement are far wider than wars and civil conflicts. Despite this, there are still no nationally or internationally binding agreements or treaties that guarantee the rights of people who have been uprooted by other causes, such as environmental disruption, disasters/catastrophes, or development projects (such as the Three Gorges Dam project in China). The *United Nations Guiding Principles on Internal Displacement* defines IDPs as:

> ... persons or groups of persons who have been forced or obliged to flee or leave their homes or places of habitual residence, in particular as a result of or in order to avoid the effects of armed conflict, situations of generalized violence, violations of human rights or natural or human made disasters, and who have not crossed an internally recognized state border (United Nations, 2008, Introduction).

However, although widely recognized as an international standard, and certainly helpful in guiding NGOs and other aid organizations in assisting internationally displaced persons, the guiding principles have not been agreed upon in a binding covenant or treaty and have no legal standing. We must also recognize the very real potential for global climate change to generate displacements and migrations across international borders.

Given the dearth of appropriate policies for IDPs, the need for developing adequate legal protections and assistance programs for populations facing potential displacement by forces generated by global climate change becomes urgent. Current estimates for the number of environmentally displaced people around the world are highly debated for reasons discussed earlier, but the Office of the United Nations High Commissioner for Refugees estimated in 2002 that approximately 24 million people around the world had been displaced by floods, famines, and other environmental causes (UNHCR, 2002, p. 12) (see Box 8.3). These numbers could be dwarfed by the potential displacements caused by future global climate change. There is a great need currently for legally binding policies and informed practice to address the massive displacement and resettlement that global climate change is projected to cause.

Vulnerability science has made clear that exposure to hazards alone does not determine where the serious impacts will most likely be experienced. The challenge lies in determining not just absolute exposed land and absolute exposed population, but specific lands and populations in different socially configured conditions of resilience or vulnerability. For example, the problem with assessing the exposure of both land and

BOX 8.3 ESTIMATING ENVIRONMENTALLY INDUCED DISPLACEMENT

UNHCR (2002): 24 million
El-Hinnawi (1985): 50 million
The Almeria Statement (1994): 135 million
Myers (2005): 200 million
The Stern Review (2006): 200 million by 2050
Nicholls (2006): 50 to 200 million by 2080
Friends of the Earth (2007): 200 million by 2050; 50 million in Africa alone
Christian Aid (2007): 250 million by climate change; 645 million by development projects

Source: FEMA EMI Higher Education Program.

population to sea level rise is that we are dealing with more than projected increases in sea level. One can look at the Dutch, who have been reasonably successful in dealing with fluctuations in sea level because it is a national priority. We must also consider various future projections about different societal and environmental trajectories including greenhouse gas emissions, demographic change, migration trends, infrastructure development, mitigation strategies, adaptive capacities, vulnerabilities and patterns of economic change, all of which will play out in different ways, according to the political, economic, and socio-cultural dispositions of national governments, international organizations, and general populations. This is all subject to the great unknown as to when catastrophes of no or little notice will reshape the political landscape.

8.12 IDENTIFYING THE POTENTIALLY DISPLACED

The concept of social vulnerability will be a key in identifying beforehand those groups that may suffer substantial displacement, but vulnerability assessment is still at an early stage of development. However, the poor and underdeveloped regions of the world are likely to have fewer resources with which to deal with climate change.

Global climate models currently lack sufficient resolution to profile types and magnitudes of changes to be expected in specific local sites. There are also real limitations in our abilities to assess human vulnerabilities to these projected, yet ill-defined, threats. The interaction of multiple, sometimes rapidly changing stresses, such as economic shocks, natural hazards, etc., with systemic chronic stresses, such as malnutrition and poor health, in shaping vulnerability and the dynamic interaction of these forces across social and geographic scales means that levels of vulnerability will continue to be hard to predict for the foreseeable future (Dow et al., 2006).

8.12.1 United Nations Framework Convention on Climate Change (UNFCCC) Article 4.8

Article 4.8 of the UNFCC provides that parties shall give full consideration to what actions are necessary ... to meet the specific needs and concerns of developing country parties arising from the adverse effects of climate change and/or the impact of the implementation of response measures, especially on: (see Box 8.4)

BOX 8.4 UNFCC ARTICLE 4.8

a. Countries with low lying coastal areas
b. Countries with arid and semiarid areas, forested areas, and areas liable to forest decay
c. Countries with areas prone to natural disasters
d. Small island countries
e. Countries with areas liable to drought and desertification
f. Countries with areas of high urban atmospheric pollution
g. Countries with areas with fragile ecosystems, including mountainous ecosystems
h. Countries whose economies are highly dependent on income generated from the production, processing, and export, and/or on consumption of fossil fuels and associated energy-intensive products
i. Land-locked and transit countries

8.13 MASS RELOCATION AS MITIGATION

Mitigation aims to increase the self-reliance of people in hazard prone environments to demonstrate that they have the resources and organization to withstand the worst effects of the hazards to which they are vulnerable. In other words, disaster/catastrophe mitigation, in contrast to dependency creating relief, is empowering. Mitigation thus brings in issues of development in that anything that increases the resilience and security of society can be defined as a form of development. At the time this book was written, the literature failed to identify any major examples of mass relocation as mitigation. There are, however, a number of examples of small-scale voluntary resettlement as mitigation that have enhanced the resilience and reduced the vulnerability of communities, including many examples from FEMA's now-defunct Project Impact program.

8.14 SOCIAL VULNERABILITY

By *vulnerability* we mean the characteristics of a person or group in terms of their capacity to anticipate, cope with, resist, and recover from the impact of a hazard. It involves a combination of factors that determine the degree to which someone's life and livelihood is put at risk by a discrete and identifiable event in nature or society (Wisner et al., 2004).

Although the problem seems to be an enduring one, until relatively recently, most of the literature has characterized disaster/catastrophe-induced displacement as temporary, suggesting that people eventually return (Oliver-Smith, 1991). However, a great deal of this research has been excessively event-focused both temporally and spatially and has given far less attention to the longer term, more geographically dispersed aspects of postevent recovery or reconstruction. Today, with broadened and deepened spatial and temporal scales of analysis, natural disasters/catastrophes, rather than unanticipated and unique events caused by a natural agent, are seen to be much more explainable in terms of the "normal" order of things. That is, the conditions of inequality and subordination in the society rather than the accidental geophysical features of a place are primary sources of vulnerability. This perspective has shifted the focus away from the disaster/catastrophic event and toward the vulnerability of peoples embedded in the "ongoing societal and man–environment relations that prefigure [catastrophe]" (Hewitt, 1983).

The concept of vulnerability refers to the totality of relationships in a given social situation producing the formation of a condition that, in combination with environmental forces, produces some form of catastrophe. Risks and outcomes are thus largely socially produced. This more complex understanding of vulnerability and catastrophes enables researchers to analyze how social systems generate the conditions that place different kinds of people, often differentiated among variables, such as class, race, ethnicity, gender, or age, at different levels of risk from the same hazard, and different forms of suffering from the same event. Vulnerability to these dynamic processes is particularly accentuated in the developing world, where people have fewer resources either to manage threats or to recover from impacts. Therefore, overwhelming as the processes of global climate change threaten to become, their impacts, like any disaster, will be socially, politically, and economically mediated, distributed, and interpreted. The measures taken to mitigate and respond will be similarly structured. As an example, more developed countries may enforce land use and building code regulations that force people to avoid building in inherently dangerous areas, such as floodplains, landslide areas, and earthquake fault rupture zones, or comply with building codes intended to minimize or avoid damages to the built environment from known hazards.

8.15 DEMOGRAPHIC MOVEMENT

BOX 8.5 DEMOGRAPHIC MOVEMENT VOCABULARY

- *Flight:* Escape from a life threatening agent
- *Evacuation:* Removal of people from harm's way
- *Migration:* People move to a new home ground either voluntarily or involuntarily in varying degrees
- *Displacement:* The purposeful involuntary uprooting of people from a home ground
- *Resettlement:* Purposeful, adaptive, or mitigative relocation of people to new sites either voluntarily or involuntarily in varying degrees

When people are subject to internal or external forces that require them to alter their location in space, the resulting movement takes a number of different forms that vary along the spectrum of a number of different characteristics. If the threat is immediate, *flight* or *escape* to the closest safe location is a frequent response (Figure 8.8). An impending threat may result in an evacuation that may resemble flight or may be more organized or administrated by internal or external agents. *Displacement* similarly can occur as the result of flight or be more planned in the sense that people are organized and obliged to move from one residence site to another either temporarily or permanently. If the movement is thought to be permanent, *resettlement* in the form of the creation of a new residence site may actually be the outcome. Finally, as mentioned earlier, forced *migration* involves permanent, longer distance moves

FIGURE 8.8 Thousands of residents of New Orleans took flight on August, 29, 2005, from the floodwaters left behind by Hurricane Katrina to high ground under Interstate 10 and await evacuation out of the city. Hundreds of buses worked throughout the day and night to carry people to a shelter (an example of *displacement*). (Photo by Win Henderson/FEMA.)

generally into very different environments. Some of the forms of demographic movement may lead to others as flight or evacuation may lead to displacement and resettlement or eventually to forced migration.

8.16 CONTINUA OF DISPLACEMENT

BOX 8.6 SOCIAL AND ENVIRONMENTAL PAIRS ASSOCIATED WITH DISPLACEMENT

- Mitigative–Adaptive
- Voluntary–Involuntary
- Temporary–Permanent
- Physical danger–Economic danger
- Administrated–Nonadministrated

Each form of demographic movement may vary along a number of scales or continua associated with certain characteristics that refer largely to the social and environmental relations expressed in the particular context. In addition, these concepts have to be treated with a certain degree of flexibility, and should not be taken as hard and fast categories because the reality of particular occurrences of forced migration tends to be too complex to classify within rigid categories.

Looking at each kind of demographic movement along the various continua presented reveals the wide variability that each can display. Flight, for example, generally tends to be proactive and forced, but not administered; can be temporary or permanent; tends to be associated with physical danger; and may result in permanent displacement, resettlement, or forced migration. Evacuation, usually a response to physical danger, can have similar outcomes, can be proactive or reactive, but tends to be administered to a greater or lesser degree.

In a catastrophe, the reader should make a distinction, one that holds for development-induced movement as well, between forced displacement and forced migration. Displacement can be an administered involuntary process of moving a population. Displacement can be temporary or permanent, voluntary or involuntary, and may be a response to both physical and economic harm. Migration involves moving farther away, to different environments, and for longer periods of time, if not permanently, and will vary in the degree of willingness on the part of the those being relocated. Except in extreme cases, the coercive power or push factors in catastrophe-induced forced migration will vary and may be balanced to some degree by pull factors or positive inducements to move.

There has been much speculation regarding the number of people who may be forced to migrate as a result of various forms of catastrophes, but surprisingly little about where they are going to go and what they are going to do when they get there. From other forms of forced migration and displacement, we can project that there will be at least four options.

1. **Assimilation with co-ethnics:** This approach sometimes works and sometimes does not. For the past 40 years, Angolan refugees who crossed into Zambia have not felt displaced because they have moved in with co-ethnics even though they had crossed a political border. Each refugee was provided a plot of land by the Zambian government. Some refugees have lived on these plots since 1971. On the other hand, native-American Inuits in Alaska can't conceive of living in someone else's community, even though they also may be Inuit. They say it would be like living in someone else's house.
2. **Camps:** These are often seen as a temporary solution for political refugees, but they have a tendency to become long term as in

the case of the Palestinians. In some cases, once a camp is established it may become a near permanent fixture, even though different refugee populations may pass through it en route to repatriation or resettlement in a foreign country as in the Tongorara camp in Zambia that has housed successive waves of different refugees from regional conflicts over a generation.

3. **Urbanization:** The vast majority of people displaced over the past 40 years by India's Narmada Valley Dam projects have migrated to urban slums, disappearing into those vast populations of poor and vulnerable people that strain the capacities of environments, administrations, and infrastructures to meet urban needs.
4. **Resettlement:** The outcomes of the vast majority of planned resettlement projects have been very poor, resulting in the impoverishment of the affected populations. Such projects are frequently poorly planned, underfunded, and badly implemented, leaving affected populations destitute and disempowered. Such projects compound the losses experienced in displacement and if managed poorly can constitute severe violations of human rights.

8.17 DISPLACEMENT AND LOSS

BOX 8.7 LOSSES ASSOCIATED WITH DISPLACEMENT

- Economic: land, tools, markets, jobs
- Social: networks, kin, clientele
- Political: power, host–guest conflict
- Cultural: place and identity

For people affected by catastrophes and other environmental changes, displacement and resettlement often constitute a second catastrophe in their lives. The complexity of a catastrophe reverberates in the losses that people experience and in the process of recovery. Catastrophes inflict terrible losses on people and communities, often shredding families and uprooting communities to radically different or new environments. Displacement both compounds and makes permanent many of the losses incurred in catastrophe. Those who can reconstruct in-place, even in much diminished circumstances, generally stand a better chance at recovery than those that are displaced.

Recovery from the destruction or loss through uprooting of livelihood and community require affected people to engage in a process of reinvention. As human beings are social creatures, the reinvention of the self will be intimately linked to the reinvention of social bonds and community as the principal form of social living of humankind. The process of reinvention or recovery will have both material and social aspects. Material and social losses compound each other. Those who are uprooted, having suffered almost complete loss, like political refugees, almost always migrate with fewer resources with which to reconstruct their lives.

Material elements, such as housing, the possessions of a lifetime, infrastructure, services like electricity and potable water, healthcare, transportation and communication, and nutrition can all be endangered, damaged, or destroyed in catastrophes or lost in subsequent displacement. In addition to physical damage, material losses resonate profoundly as well in the social world, compounding the serious losses also inflicted in the economic, social, and cultural life of survivors. For example, material damages frequently mean the loss of livelihoods, whether through destruction or loss of worksite, tools, and equipment; land or common property resources; or physical injury. Loss of livelihood and the capacity to sustain oneself endanger individual and social identity, producing a loss of status and resulting in marginalization and social disarticulation. The loss of a house is also the loss of a place in your social world order. And the community, the social world, is endangered by such individual losses.

The dispersal, displacement, or death of family members fragment not only a household, but erodes the social cohesion of a community as well. Catastrophe-caused deaths shred those networks of relationships that form the basis of personal and social identity, tearing the social fabric, setting people adrift, without those ties that anchor the self in the social world. Survivors of catastrophes, in which there is great loss of life and prolonged devastation and displacement, also may suffer a loss of personal identity, the partial loss of the self. The loss of significant others in high mortality catastrophes is also a loss of the self in that the part of the person that was invested in the lost relationship is also lost. Thus, the loss of a child means that one has lost that part of the self that was a parent. The loss of status, the resulting social leveling, and the reduction to a common level of misery all can constitute an assault on the sense of the self.

Cultural identity is often placed at risk in uprooted communities. The loss and destruction of important cultural sites, shrines, religious objects, the interruption of important sacred and secular events and rituals undermines the community's sense of itself. One of the few positive outcomes of Hurricane Katrina's assault on New Orleans was that the city's cultural identity was not lost since the French Quarter and famed Garden Distinct were spared the ravages of flooding. It should not be lost on the reader that the original settlers of New Orleans knew to build on the high ground consisting of sand bars deposited by the Mississippi River. Had they flooded, the recovery of the city would be much more difficult to envision in a positive light. Catastrophes and displacement may endanger the identification with an environment that may once have been seen as nurturing and central to cultural identity, but is feared and distrusted in the aftermath (Oliver-Smith, 1992). Displacement for any group can be a crushing blow, but for indigenous peoples it can prove mortal. Retainage of land rights is considered to be an essential element in the survival of indigenous societies and distinctive cultural identities.

These losses of community, family, and self compound each other to often create another form of loss: the loss of meaning. These events and the prolonged conditions of deprivation and displacement can shake the foundations of personal worldview and identity. They challenge the culturally constructed vision in which the world is a logical and just place, where life makes sense. Catastrophes and displacement rob people of the social context in which they have lived meaningful lives to the present, and were judged to be significant by others around them. This loss of personal relationships and the social context in which they were expressed and in which the individual was affirmed, may leave people devoid of a sense of meaning, a sense of purpose in life. Religious belief can also become a casualty in the aftermath of catastrophes as people may question a god's providence in allowing such an event to occur.

When people are forced from their known environments, they become separated from the material and cultural resource base upon which they have depended for life as individuals and as communities. Moreover, a sense of place has been shown to play an important role in individual and collective identity formation, in the way time and history are encoded and contextualized, and in interpersonal, community, and intercultural relations (Altman and Low, 1992; Rodman, 1992; Escobar, 2001). A sense of place is crucial in the creation of what Giddens (1990, p. 102) calls an "environment of trust" in which space, kin relations, local communities, and tradition are linked (also cited in Rodman, 1992, p. 648).

In summary, removal from one of the most basic physical dimensions of life can be a form of removal from life. The disruption in individual or community identity and stability in place, resulting in resettlement in a strange landscape, can baffle and silence people in the same way a strange language can (Basso, 1988; also cited in Rodman, 1992, p. 647). Culture loses its ontological grounding (nature of being) and people must struggle to construct a life world that can clearly articulate their continuity and identity as a community again. The human need for "environments of trust" is fundamental to the sense of order and predictability implied by culture.

8.18 INVOLUNTARY DISPLACEMENT AND RECOVERY

If, in fact, the uprooted are resettled in some systematic way, the quality of the resettlement project itself may play a major role in the capacity of the community to recover from the trauma of displacement. Such projects are really about reconstructing communities after they have been materially destroyed and socially traumatized to varying degrees. Reconstructing and reconstituting a community is an idea that needs to be approached with a certain humility and realism about the limits of our capacities. Such humility and realism have not always characterized the planners and administrators of projects dealing with uprooted peoples to any major extent to date. Indeed, the goals of such undertakings frequently stress efficiency and cost containment over restoration of a community. Such top-down initiatives have a poor record of success because of a lack of regard for local community resources. Planners often perceive the culture of uprooted people as an obstacle to success rather than as a resource worth preserving.

Reconstructing/reconstituting a community means attempting to replace through administrative efforts an evolutionary process in which social, cultural, economic, and environmental interactions arrived at through trial and error and deep experiential knowledge develop, enabling a population to achieve a mutually sustaining social coherence and material sustenance over time. The systems that develop are not perfect, are often far from egalitarian, and do not conform to some imagined standard of efficiency. The idea that such a process could be the outcome of planning is ambitious to say the least. One of the best outcomes that might be imagined for resettlement projects is to work out a system in which people can materially sustain themselves while they themselves begin the process of social reconstruction. The least that could be hoped for might be that resettlement projects not impede the process of community reconstitution. However, if the level of impoverishment experienced by most resettled peoples is any indicator, even adequate systems of material reproduction are beyond either the will or the capabilities of most contemporary policy makers and planners. This does not bode well for the victims of potential mass displacements.

This isn't to say that there can't be positive outcomes from the relocation process. One positive impact might be that the relocation site is safer than the original settlement site. Examples might include relocation from a riverine floodplain, coastal high hazard area, or from an earthquake surface rupture or soil liquefaction zone. Another benefit might be that a relocation site offers more economic development opportunities, such as relocating an agricultural-based community to an area with more fertile land.

8.19 RESETTLEMENT

Over the last half-century, various researchers have developed a variety of conceptual approaches to the problem of mass relocation. First, Scudder (1981) alone and, subsequently, with Colson, developed an approach based on the concept of stress to describe and analyze the process of involuntary dislocation and resettlement (1982). "The Four Stage Framework," as Scudder (2009) now calls it, emphasizes how most resettlers can be expected to behave during each of the four stages, passage through which must be completed if the resettlement project is to be successful.

They posited that **three forms of stress** resulted from involuntary relocation and resettlement: physiological, psychological, and socio-cultural. These three forms of stress, referred to as multidimensional stress, are experienced as affected people pass through the displacement and resettlement process.

Physiological stress is seen in increased morbidity and mortality rates. "Dying of a broken heart" with prevalence among the elderly. Psychological stress, seen as directly proportional to the abruptness of the relocation, has four manifestations: trauma, guilt, grief, and anxiety:

- *Trauma* from the uprooting process
- *Guilt* about having survived
- *Grief*, the "lost home" syndrome
- *Anxiety* about an uncertain future

Socio-cultural stress is manifested as a result of the economic, political, and cultural effects of relocation.

- Lack of prepared *economic* support (i.e., houses instead of means of livelihood)
- *Political* leadership vacuum (former leaders who failed to protect or resist; problems with the host community)
- Reduction in *cultural* inventory: loss of traditional patterns, institutions, and symbols; conflict with the host community

The resettlement process itself is represented as occurring in four stages, which they label:

1. Recruitment
2. Transition
3. Potential development
4. Handing over/incorporation

Recruitment refers to the decisions taken by authorities regarding the population to be relocated, particularly those that influence the length and severity of the stressful transition stage. The transition stage begins when the population to be relocated is first affected. Generally speaking the transition stage is the longest and the stage in which the most severe multidimensional stress is experienced. The general attitude of people during the transition stage is conservative in order to avoid the possibility of further risk and stress. The stage of potential development begins when people begin to abandon their conservative risk avoidance strategies and express greater initiative and risk-taking behavior. Scudder and Colson (1982) emphasize that this stage is often never realized because many projects remain trapped in the transition stage by inept and inappropriate policy and implementation. Equally difficult to attain is the final stage of handing over or incorporation. Achieving the incorporation stage signifies that the resettlement project has been successful. They define success by the act that occurs when higher authorities hand over economic and political affairs to local management and phase out external agencies and personnel from day-to-day management of the community. The community has become able to assume its place within the larger regional context that includes host communities and other regional systems.

BOX 8.8 FOUR KEY FACTORS IN SUCCESS/FAILURE OF POSTDISASTER RESETTLEMENT

- **Site selection:** proximity to resources, livelihoods, employment
- **Settlement design:** appropriate material context for social and cultural interaction
- **Housing:** culturally appropriate size/design/spacing and materials
- **Participation:** Planning with people rather than for people

Poor choice of site for resettlement is one of the most frequently mentioned causes of resettlement failure. Sites for resettlement are often chosen with factors other than the welfare and development of the population in mind.

The design or layout of the settlement has been cited as a source of sufficient dissatisfaction with resettlement to result in abandoning the site. Ease of construction, misconstrued concepts of efficiency, and the imposition of urban middle class values on rural populations seem to lie at the root of problems of monotonous, uniform designs for resettled populations that fail to consider cultural and other social contexts of the relocated population.

Housing design and reconstruction are often blamed for the rejection or failure of resettlement projects. Faulty construction and inferior materials in houses, due to poor project oversight, become quickly evident with use and create difficult living conditions, particularly regarding thermal protection in different seasons.

Projects that suffer failure or at best partial success are often characterized by policies that depend very little on consultation with the population being resettled. The three previously mentioned issues regarding poor site selection, inappropriate design, and unsatisfactory housing derive from a lack of consultation with and participation by the affected people. This lack is generally due to a disparagement of local knowledge and culture on the part of policy makers and planners.

In material terms, the needs of individuals, households, communities, and the external systems of which they are parts and the organized responses to these needs are numerous, diverse, and interconnected. While there are urgent needs in any uprooting crisis, relatively adequate procedures have actually been developed to respond to these, although a uniform standard has yet to be reached, despite the much-debated Sphere Project guidelines for reaching such standards (2000). Unfortunately, the procedures put in place to cope with emergency needs are rarely linked to key features of community organization, although they can have a determining role in the development of the longer term rehabilitative system, often with very negative impacts on the long-term viability of the community.

Homes and life-sustaining activities are the most deeply felt needs in establishing a long-term system for dealing with material necessity in the stress of uprooting and resettlement, whatever the cause. Whether it is due to sudden onset of a catastrophe, the explosion of civil violence, or the bad or absent planning of development projects, resettled people are frequently housed in temporary quarters, which in all too many cases become permanent, however inadequate and inappropriate they may be. Housing and settlement design that are donor-driven endanger the connection that people establish with their built environment, violating cultural norms of space and place, inhibiting the reweaving of social networks and inhibiting the reemergence of community identity

(Oliver-Smith, 1991). Additionally, the design of settlements that place living quarters in dense proximity, which is convenient for relief agencies, significantly increases the probability of infectious disease transmission within the community. (Toole, 1997)

The other great need to be addressed at the material level for the uprooted is employment. From both a material and a psychological standpoint, economics drives the process of reconstruction and resettlement. Employment provides needed income to replace or improve upon those personal and household needs not provided by aid, but it is also a form of action that enables people to return to being actors rather than being acted upon as disaster victims or refugees, all of which are essentially passive rather than active societal roles. Uprooting causes many people to lose their means of production, whether it is the land, tools, or access to other resources, and they will be unable to resume normal activities until such resources are obtained. There may be a difficult trade-off between reconstituting economic resources (especially land and property) and the social and cultural benefits gained by staying together. This is especially true in development-induced displacement when a project has opted for land-for-land replacement and the host population is dense. It may be difficult or even impossible to settle a community together on sufficient land. Or people may need to move far from community networks in order to have both sufficient land and avoid dispersal of the community itself. This creates hard choices for the relocated. However, until people resume employment, they remain dependent on external resources and reconstruction remains incomplete.

BOX 8.9 COUNTERING RESETTLEMENT RISKS THROUGH THE IMPOVERISHMENT RISKS AND RECONSTRUCTION (IRR) MODEL

From Landlessness	to	Land-Based Resettlement
From Joblessness	to	Reemployment
From Homelessness	to	House Reconstruction
From Disarticulation	to	Community Reconstitution
From Marginalization	to	Social Inclusion
From Expropriation (loss through catastrophe)	to	Restoration of Community Assets/Services
From Food Insecurity	to	Adequate Nutrition
From Increased Morbidity	to	Better Healthcare

8.20 IMPOVERISHMENT RISKS AND RECONSTRUCTION MODEL

BOX 8.10 EIGHT BASIC RISKS OF DISPLACEMENT AND RESETTLEMENT

1. Landlessness
2. Joblessness
3. Homelessness
4. Marginalization
5. Social Disarticulation
6. Food insecurity
7. Increased morbidity
8. Loss of access to common property resources

At roughly the same time that Scudder and Colson (1982) were developing their model in the early 1980s, an approach began to gain favor that focused on the linked ideas of vulnerability and risk. Vulnerability was initially employed in disaster/catastrophe research to understand the vast differences among societies in disaster/catastrophe losses from similar agents. An alternative perspective on human–environment relations, emphasizing the role of human interventions in generating disaster/catastrophe risk and impact, found that these sets of relations coalesced in the concept of vulnerability (Hewitt, 1983).

As these concepts gained currency, Cernea (1990) began to write about the risks of poverty resulting from displacement from water resource projects. He subsequently developed his now well-known Impoverishment Risks and Reconstruction (IRR) approach to understanding (and mitigating) the major adverse effects of displacement in which he outlines eight basic risks to which people are subjected by displacement (1996; Cernea and McDowell, 2000). The model is based on the three basic concepts of risk, impoverishment, and reconstruction. Deriving his understanding of risk from Giddens' (1990) notion of the possibility that a certain course of action may produce negative effects, Cernea models displacement risks by deconstructing the multifaceted process of displacement into its identifiable principle and increased morbidity, loss of access to common property resources, and social disarticulation (Cernea, 2000). He further asserts that the probability of these risks producing serious consequences is extremely high in a badly or unplanned resettlement. All these risks follow the displacement process with the threat of a second calamity that entails such risks that can translate directly into losses. Cernea's IRR model is designed to predict, diagnose, and resolve the problems associated with development-induced displacement and resettlement.

8.20.1 Displacement and Sustainable Livelihoods

Two years later, McDowell (2002) combined *sustainable livelihoods* research and Cernea's IRR approach to develop a methodological framework for research on postdisaster/catastrophe resettlement. His approach is based on the assertion in sustainable livelihoods research that social institutional processes are central to livelihoods in the ways they influence households' access to resources, whether natural or social. One of the principal risks in displacement and resettlement is social displacement, including the scattering of kinship groups and informal networks of mutual help (Cernea, 2000). The disarticulation of spatially and culturally based patterns of self-organization, social interaction, and reciprocity constitutes a loss of essential social ties that affect access to resources, compounding the loss of natural and humanmade capital. Thus, in displacement and resettlement, people's adaptations to the social displacement produce new dynamics that influence their access and control over resources, often leading to a process of further impoverishment. Therefore, understanding institutional processes in resettlers' adaptive strategies will be crucial for identifying the socio-culturally specific nature of the risks Cernea identified as inherent in forced displacement, thus helping to explain why displacement and resettlement so often result in greater impoverishment of affected households.

8.20.2 Displacement and Resettlement as a Complex System

More recently, de Wet (2006) has sought to incorporate Cernea's important insights into a more comprehensive approach. Asking why resettlement so often goes wrong, de Wet sees two broad approaches to responding to the question. The first approach is what he calls the "Inadequate Inputs" approach, which argues that resettlement projects fail because of a lack of appropriate inputs: national legal frameworks and policies, political will, funding, predisplacement research, careful implementation, and monitoring. Optimistic in tenor, the inadequate inputs approach argues that the risks and injuries of resettlement can be controlled and mitigated by appropriate policies and practices. On the other hand, de Wet finds himself moving toward what he calls the "Inherent Complexity" approach. He argues that there is a complexity in resettlement that is inherent in "the interrelatedness of a range of factors of different orders: cultural, social, environmental, economic, institutional, and political—all of which are taking place in the context of imposed space change and of local level responses and initiatives" (de Wet, 2006). Moreover, these changes are taking place simultaneously in an interlinked and mutually influencing process of transformation. And further, these internal changes from the displacement process are also influenced by and respond to the imposition from external sources of power as well as the initiatives of local actors. Therefore, the resettlement process emerges out of the complex interaction of all these factors in ways that are not predictable and that do not seem amenable to a linear-based, rational planning approach.

de Wet suggests that a more comprehensive and open-ended approach rather than the predominantly economic and operational perspective of the inadequate inputs approach is necessary to understand, adapt to, and take advantage of the opportunities presented by the inherent complexity of the displacement and resettlement process. While some might see this perspective as unduly pessimistic, the fact that authorities are limited in the degree of control they can exercise over a project creates a space for resettlers to take greater control over the process. The challenge thus becomes the development of policy that supports a genuine participatory and open-ended approach to resettlement planning and decision making (de Wet, 2006).

8.20.3 Resettlement, Reconstruction, and Development

Resettlement must be approached as development. If mass relocation results only in the dispersal of affected populations to poverty stricken slums, or warehousing them in temporary or otherwise permanent camps (rural slums), the process of resettlement will compound the trauma and human rights violations of uprooting and consign them forever to misery. Moreover, after 40 years of application in development-forced displacement, the compensation principle has been amply demonstrated to be utterly inadequate in restoring livelihoods to displaced people (Cernea and Oliver-Smith, 2009; Scudder, 2009). Therefore, resettlement projects must be configured as development projects, with the appropriate investments to enable people to become active and self-sufficient members of resilient communities.

8.21 RESPONSIBLE AGENCIES IN MASS RELOCATION

Relatively few nations have either the necessary legislation or the administrative structure and capacity to adequately address the task of resettling displaced populations. Generally speaking, an amalgam of public agencies, with jurisdiction over a wide spectrum of environmental, social, regulatory, and economic domains, is needed to plan and implement resettlement, often producing projects that demonstrate their conflicting and often contradictory agendas.

For example, in the United States 33 federal, state, and county agencies were involved in the relocation of the community of Allenville, Arizona, away from a floodplain. Although this comparatively small relocation was successful, the project was characterized by, at times, bewildering complexity in which the various rules and operating procedures of the agencies involved became a major impediment to the successful resettlement of the population (Perry and Mushkatel, 1984).

A study by the World Bank's Operations Evaluation Department (OED) of five major bank-funded dam projects concludes that while better planning has occurred, it has not led generally to better involuntary resettlement. Furthermore, the public agencies charged with resettlement have not responded adequately to the challenge of resettlement. They also find that income restoration strategies, whether based on land-for-land or other options, have not in general been successful. The key to success, in their opinion, is genuine commitment to the resettlement process as a development opportunity by the borrower country (Picciotto et al., 2001).

Resettlement Action Plans (RAPs) are now a requirement of multilateral lending institutions, such as the World Bank, the Inter-American Development Bank, and the Asian Development Bank, prior to undertaking the resettlement of any population for the construction of an infrastructural project.

8.22 RESETTLEMENT ACTION PLAN

BOX 8.11 SPECIFIC PLANNING PRINCIPLES OF RAPS

- Policy framework
- Income enhancement
- Entitlements
- Consultation
- Strengthening institutional capacity
- Resettlement budget

Not all nations have a national involuntary resettlement policy and, of those that do, most are inadequate. Income enhancement is rarely accomplished, resulting in the policy objectives not being realized. In development projects, not all those displaced are equally affected and, thus, may be accorded different benefits and entitlements that can lead to a sense of inequity of the part of the displaced. Consultation constitutes a big debate between those who favor *prior-informed consultation* and those espousing *prior-informed consent*, usually resulting in a huge failing in development-forced resettlement. Projects are unpopular, often career killers for those leading the effort, based on poor research and poor implementation. Resettlement projects are almost always underfunded and compensation packages unequal to the task of restoring livelihoods and well-being.

BOX 8.12 STANDARD OUTLINE FOR A RAP

Income restoration	Income restoration
Efforts to minimize resettlement	Institutional arrangements
Census and socioeconomic survey	Implementation schedule
Introductory description of project	Participation and consultation
Legal framework	Grievance redress
Resettlement sites	Monitoring and evaluation
Introductory description of project	Costs and budgets
Efforts to minimize resettlement	Annexes
Census and socioeconomic surveys	

8.22.1 Example of Success: Preparatory Stages for the Arenal Dam Resettlement Project

The Arenal Hydroelectric Project in Costa Rica is considered by many to have succeeded in improving the standards of living and returning control over their own lives to the resettled people five years after the implementation of the project (Partridge, 1993). The Arenal Hydroelectric Project involved the construction of a dam 70 meters (nearly 230 feet) high that would produce a reservoir of 1,750 cubic meters that necessitated the displacement and resettlement of about 2,500 people (roughly 500 families).

The area in which the project was located was characterized by the "humid tropics cattle-ranching complex," with little or no commercial agriculture. Subsistence agriculture was declining as well. When the dam project was approved, resettlement planning began two years before actual construction began. The preparatory period consisted of 11 steps or phases (see Box 13).

BOX 8.13 THE 11 PREPARATORY PHASES OF THE ARENAL DAM SETTLEMENT PROJECT

Phase I: Ethnographic Sample Survey of Communities
Phase II: Information Campaign and Meetings with Families
Phase III: Census of People and Property to be Affected
Phase IV: Making Public the Planning Data
Phase V: New Settlement Site Selection
Phase VI: Action Plan for Resettlement Prepared
Phase VII: Land Acquisition
Phase VIII: Participation of the Affected Population
Phase IX: Financial Mechanism for Restitution of Property
Phase X: Construction of New Settlements
Phase XI: Community and Agricultural Development

The first two to three years (1976 to 1979) of the new communities were difficult. The resettled people cleared land and planted traditional crops, such as maize, manioc, plantains, and bananas. Seeds and cuttings were made available to the farmers as soon as their new plots had been allocated. The agricultural system in the new settlements was initially the traditional slash and burn technology. Subsequently, new vegetable, tree, and pasture crops, which had been field tested during the construction phase, began to be cultivated, particularly a new variety of coffee. Individual farmers obtained loans from the National Bank of Costa Rica to intensify production, and the farmers as a group organized a marketing cooperative. The income derived from growing coffee increased by roughly 100% over preresettlement levels. New grasses for cattle fodder also enabled farmers to increase cattle pastured from one animal per hectare to three. A new road constructed by the project linked the new communities to market centers and fostered the development of several small dairy farms.

Income from these initiatives stimulated the purchase of additional farmlands, the construction of outbuildings on farms, purchase of vehicles, and the construction of a rural school building with no assistance from the government. The success of the families in the farming sector nourished success in the commercial sector. Shopkeepers and their families benefited by the increased levels of cash income in the communities. In the new communities, the levels of fixed capital and inventory values ranged between 50% and 200% greater than in the old settlements. Furthermore, social organizational features developed in the new communities in the form of a school committee, a sports committee, and the continuation of the Catholic Church committee.

Partridge (1993) attributes the successes achieved in the project to three basic steps in the preparation process. Good data collection and community studies carried out by social scientists resulted in a resettlement plan that was both realistic and practical. He also emphasized the importance of consultation with the people to be relocated and their meaningful participation in the preparation process. Also significant in the success of Arenal was the strategy of establishing the new farms on the basis of traditional crops and technology, allowing the farmers to continue with known practices for the initial period of adjustment. After three years, innovations were introduced when people were more able to assume risks, and the results significantly enhanced income generation, leading to long-term acceptance of the new settlements.

8.23 THE NEAR FUTURE

Global climate change is expanding threats and hazards, and expanding vulnerable populations are occupying hazardous locations, creating high potential for mass population displacements. Wars and civil conflicts diminished slightly in the early years of the century, but are now increasing the number of refugees again as a result of the political destabilization of Africa from decolonization and Eastern Europe from the collapse of the Soviet Union. Development agendas are shifting back to high risk, high gain, large infrastructural projects. While future focus on vulnerability reduction and mitigation, as well as adaptive capacity and resilience building, is needed, there is also need to focus on reconstruction, relocation, and resettlement both as mitigation and reconstruction.

8.24 DISCUSSION QUESTIONS

1. Identify any recent trends and patterns of catastrophe-forced displacement and resettlement.
2. What are the specific forms of social vulnerability that make mass relocation, from a catastrophe, probable?
3. How will the policy discourse and practice of institutional players (states, international development, and aid agencies) frame, define, and categorize catastrophe-forced displacement and resettlement?
4. What is the linkage between vulnerability and rights/entitlements and the capacity to reconstruct livelihoods?
5. What is the relationship between the *material* and the *social* in mass relocation?
6. Discuss the RAP outline and its possible application to situations of catastrophic driven mass relocation.

REFERENCES

Adger, W.N., J. Paavola, and S. Huq. 2006. Toward justice in adaptation to climate change. In *Fairness in adaptation to climate change*, eds. W. Neil Adger, J. Paavola, S. Huq, and M.J. Mace, pp. 1–20, Cambridge, MA: MIT Press.

Altman, I. and S. Low. 1992. *Place attachment. Volume 8. Human behavior and environment: Advances in theory and research*. New York: Plenum Press.

Barry, J.M. 1997. *Rising tide: The great Mississippi flood of 1927 and how it changed America*. New York: Simon and Schuster.

Basso, K. 1988. Speaking with names: Language and landscape among the western Apaches. *Cultural Anthropology* 3(2): 99–130.

Black, R. 2001. Environmental refugees: Myth or reality? UNHCR Working Papers (34):1–19.

Canclini, N.G. 1993. *Transforming modernity: popular culture in Mexico* (Translated by Lidia Lozano.) Austin: University of Texas Press.

Castles, S. 2002. Environmental change and forced migration: Making sense of the debate. UNHCR Working Papers (70): 1–14.

Cernea, M. 1990. Internal refugee flows and development-induced population displacement. *Journal of Refugee Studies*, 3(4): 320–339.

Cernea, M. 1996. *Bridging the research divide: Studying refugees and development oustees. In search of cool ground: War, flight & homecoming in Northeast Africa*, ed. Tim Alan. Geneva: United Nations Research Institute for Social Development.

Cernea, M. 2000. Risks, safeguards, and reconstruction: A model for population displacement and resettlement. In *Risks and reconstruction*, eds. M. Cernea and C. McDowell. Washington, D.C.: The World Bank.

Cernea, M. and C. McDowell, eds. 2000. *Risks and reconstruction: Experiences of resettlers and refugees*. Washington D.C.: The World Bank.

Cernea, M. and A. Oliver-Smith, eds. 2009. Financing for development: Benefit-sharing mechanisms in population resettlement. *Development and Dispossession: The Anthropology of Displacement and Resettlement*, pp. 49–76. Santa Fe: School for Advanced Research Press.

Deng, F. and R. Cohen. 1999. Masses in flight: The global crisis of internal displacement and the forsaken people: Case studies of the internally displaced. *Human Rights Quarterly* 21(2): 541–544.

de Wet, C. 2006. Risk, complexity and local initiatives in forced resettlement outcomes. In *Development-induced displacement*, ed. C. de Wet, pp. 180–202. New York: Berhahn Books.

Dow, K., R. Kasperson, and M. Bohn. 2006. Exploring the social justice implications of adaptation and vulnerability. In *Fairness in adaptation to climate change*, eds. W.N. Adger, J. Paavola, S. Huq, and M.J. Mace, pp. 79–97. Cambridge, MA: MIT Press.

El-Hinnawi, E. 1985. *Environmental refugees*. Nairobi, Kenya: United Nations Environmental Programme.

Escobar, A. 2001. Culture sits in places: Reflections on globalism and subaltern strategies of localization. *Political Geography* 20: 139–174.

Giddens, A. 1990. *The consequences of modernity*, Cambridge, MA: Polity Press.

Hansen, A. 2005. Black and white and the other: International immigration and change in metropolitan Atlanta. In *Beyond the gateway: Immigrants in a changing America*, eds. E.M Gozdziak and S.F. Martin, pp. 87–109. Lanham, MD: Lexington Books.

Hansen, A. and A. Oliver-Smith. 1982. *Involuntary migration and resettlement: The problems and responses of dislocated peoples*. Boulder, CO: Westview Press.

Hewitt, K. 1983. *Interpretations of calamity*. Winchester, MA: Allen & Unwin, Inc.

Huq, S., S. Kovats, H. Reid, and D. Satterthwaite. 2007. Editorial: Reducing risks to cities from disasters and climate change. (editorial) *Environment and Urbanization* 19(3): 3–15.

International Federation of Red Cross and Red Crescent Societies. 2004. *World Disasters Report*. Geneva: IFRC.

IPCC. 2007. Summary for policymakers. In *Climate change 2007: The physical science basis*. Contribution of Working Group I to the Fourth Assessment Report of the Intergovernmental Panel on Climate Change, eds. S. Soomon, C. Qin, M. Manning, Z. Cher, M. Marquis, K.B. Avery, M. Tignor, and H.L. Miller. Cambridge, U.K.: Cambridge University Press.

Koser, K. 2007. The global IDP situation in a changing humanitarian context. Paper presented at the UNICEF Global Workshop on IDPs, September 4. Brookings Institute, Washington, D.C.

Martin, S.F., P. Weiss Fagen, K. Jorgensen, L. Mann-Bondat, and A. Schoenholz. 2005. *The uprooted: Improving humanitarian responses to forced migration*. Lanham, MD: Lexington Books.

McDowell, C. 2001. Involuntary resettlement, impoverishment risks, and sustainable livelihoods. The Australasian Journal of Disaster and Trauma Studies, 2002-2 (http://www.massey.ac.nz/~trauma/issues/2002-2/mcdowell.htm) (accessed January 10, 2010).

McGranahan, G., D. Balk, and B. Anderson. 2007. *The rising tide: Assessing the risks of climate change and human settlements in low elevation coastal zones, environment and urbanization.* Thousand Oaks, CA: Sage Publications.

Myers, N. 1997. Environmental refugees. *Population and Environment* 19(2): 167–82.

Oliver-Smith, A. 2006. Displacement, resistance and the critique of development: From the grass roots to the global. In *Improving outcomes in development-induced displacement and resettlement,* ed. C. de Wet. Oxford, U.K.: Berghahn Books.

Oliver-Smith, A. 1991. Success and failures in post-disaster resettlement. *Disasters* 15(1):12–24.

Oliver-Smith, A. 1992. The martyred city: Death and rebirth in the Andes. Prospect Heights, IL: Waveland Press.

Oliver-Smith, A. 2002. Theorizing disasters: Nature, culture, power. In *Culture and catastrophe: The anthropology of disaster*, eds. S.M. Hoffman and A. Oliver-Smith. Santa Fe, NM: The School of American Research Press.

Partridge, W. 1993. Successful involuntary resettlement: Lessons from the Costa Rican Arenal Hydroelectric Project. In *Anthropology and involuntary resettlement: Policy, practice and theory*, eds. S. Guggenheim and M. Cernea. Boulder, CO: Westview Press.

Pearl, J. 2000. *Causality: Models, reasoning, and inference.* Cambridge, MA: Cambridge University Press.

Perry, R. W. and A.H. Mushkatel. 1984. *Disaster management: Warning, response and community relocation.* Westport, CT and London, U.K.: Quorum Books.

Picciotto, R., W. van Wicklin, and E. Rice. 2001. *Involuntary resettlement: Comparative perspectives.* World Bank Series on Evaluation and Development, Vol. 2. Washington D.C.: The World Bank.

Renaud, F.G., J.J. Bogardi, O. Dun, and K. Warner. 2007. Control, adapt or flee: How to face environmental migration? InterSecTions No. 5/2007. United Nations University Institute for Environment and Human Security.

Rodman, M.C. 1992. Empowering place: Multilocality and multivocality. *American Anthropologist* 94(3): 640–656.

Scudder, T. 1981. What it means to be dammed: The anthropology of large-scale development projects in the tropics and sub-tropics. Engineering and Science 54(4): 9–15.

Scudder, T. 2009. Resettlement theory and the Kariba case: An anthropology of resettlement. In *Development and dispossession: The crisis of development forced displacement and resettlement*, ed. A. Oliver-Smith. Santa Fe and London: SAR Press and James Currey.

Scudder, T. and E. Colson. 1982. From welfare to development: A conceptual framework for the analysis of dislocated people. In *Involuntary migration and resettlement*, eds. A. Hansen and A. Oliver-Smith. Boulder, CO: Westview Press.

Sphere Project. 2000. Online at: http://www.sphereproject.org/ (accessed January 10, 2010).

Spirtes, P., C. Glymour, and R. Scheines. 2000. *Causation, prediction and search*. Cambridge, MA: MIT Press.

Toole, M.J. 1997. Communicable disease and disease control. In *The public health consequences of disasters,* ed. E.K. Noji. New York: Oxford University Press.

Turton, D. 2003. Refugees and 'other forced migrants': Towards a unitary study of forced migration. Paper presented at the Workshop on Settlement and Resettlement in Ethiopia, January 28–30, Addis Ababa, Ethiopia.

United National High Commissioner on Refugees. 2002. Report of the United Nations High Commissioner for Refugees. Geneva: UNHCR.

United Nations High Commissioner for Refugees. 2008. United Nations guiding principles on internal displacement. Online at: http://www.unhchr.ch/Huridocda/Huridoca.nsf/0/d2e008c61b70263ec125661e0036f36e?Opendocument (accessed January 10, 2010).

Wisner, B., P. Blaikie, T. Cannon, and I. Davis. 2004. *At risk: Natural hazards, people's vulnerability and disasters*, 2nd ed. London: Routledge.

Wood, W.B. 2001. Ecomigration: Linkages between environmental change and migration. In *Global migrants, global refugees*, eds. A.R. Zolber and P.M. Benda, pp. 42–61. New York: Berghahn.

III Conclusion

Many of the same lessons learned from the research on disaster events, concerning the postcatastrophic operational environment that an emergency manager might experience, apply to catastrophic scenarios. While the need to provide assistance may be great in a catastrophe, we should not ignore the social and cultural settings in which relief is provided. The contrast between these settings and those with which we are most familiar may be most stark when providing international assistance. At the same time, communities are diverse and there may be unique considerations even within the communities in which we live.

With respect to donations management, it is not uncommon to hear of winter clothing donated to hot climates, baby formula donated to countries accustomed to nursing, and meat-based products donated to primarily vegetarian countries. Sometimes donated goods are simply dumped on the side of the road by hired drivers who refuse to wait for extended periods of time to be unloaded. Effective donations management must be an integral part of logistic management in the postcatastrophe environment.

If the goal of humanitarian assistance is truly to help those in need, it is not enough to assume that any help is better than no help at all. If we want our assistance to be helpful, it is worthwhile to mobilize resources that take social norms, values, and local conditions into account. This is why it is so critical to work with those organizations that have a long history of involvement in the impacted communities.

In Section III, we also explored critical infrastructure/key resources (CI/KR) and CI/KR protection. As a result of reading Chapter 8, the reader should appreciate that both CI/KR protection and restoration are shared responsibilities for which federal, state, and local government entities must work closely together and coordinate with the private sector. Therefore, the process of protecting CI/KR and the process of restoring CI/KR postcatastrophe is not always seamless, since different stakeholders can have varying approaches or priorities.

Any treatment of postcatastrophe CI/KR issues also has to deal with restoration of the various damaged CI/KR sectors. Unfortunately, as is true in the aftermath of ordinary disasters, there will not be sufficient resources to restore all sectors at the same time after a catastrophe. Restoration, therefore, will require triage (e.g., restoration of some

CI/KR sectors will have to be prioritized while restoration of other sectors will suffer as a result). As with protection and response, different agencies, companies, and individuals will have different ideas of what should be restored first and, in the absence of advanced planning, lack of a uniform agenda unfortunately may become the norm. Therefore, preevent planning is essential to ensure that the most good is done for the most survivors in the shortest amount of time when it comes to restoring CI/KR (remember our discussion concerning ethics at this point).

It may be easy to quickly determine that certain CI/KR sectors require immediate attention for restoration (e.g., Agriculture and Food, Emergency Services, and Water). It also may be easy to decide that restoration of others can wait for some time (e.g., National Monuments and Icons), other than for emergent tasks necessary to prevent further damage. However, the vast majority of CI/KR structures do not readily lend themselves to rapid determination of restoration priority. Can the Postal and Shipping sector be delayed for some time? Can we forego restoration of the energy sector for several weeks? If we allow restoration of commercial facilities to be significantly delayed, how many businesses will survive? Without rapid restoration of CI/KR, the impacted area begins to spiral out of control, with businesses failing or at least leaving the area and people relocating permanently in search of stable employment.

In all catastrophes, whether or not the incident is caused by a health phenomenon like a pandemic, the event will cause significant threat or actual damage to the affected population at the very time that people need their very best physical, mental, and emotional performance capability to deal with losses and move productively toward stabilization and some kind of recovery. The role of public health in this context must not be underestimated; it plays a direct role in helping people survive and recover from physical injuries, and a crucial role in helping both individuals and communities marshal their resources for the daunting tasks ahead in reconstruction and recovery. Public health also plays an important role in protecting the health of emergency response personnel. The reader would do well to consider public health issues throughout the remaining chapters, so that one starts integrating the public health viewpoint into their emergency management thinking with respect to catastrophic planning and response.

In Section III of this book, we also explored the field of displacement and resettlement research, focusing on the development of conceptual approaches, policy positions, and practice problems in the various forms of displacement and resettlement associated with catastrophic events. The range of forms that catastrophic-forced displacement and resettlement are projected to take were considered. Emphasis was placed on developing an understanding of the factors that generate both the short- and long-term risks and consequences in major dislocations, deriving understanding from data and perspectives from other forms of displacement and resettlement, including conflict and development-caused relocations. We also identified and analyzed the key components of resettlement planning as developed for infrastructure projects, assessing their utility for crafting appropriate standards and strategies for potential future mass relocation.

We also learned that mass relocation after a catastrophe, involving the physical displacement and resettlement of people, is an extremely complex process, which if not properly planned and managed (with the full participation of affected people) may result in long-term hardship for the displaced as well as potential conflict with host populations and environmental damage in locations in which they are resettled.

Over the past half-century, researchers on development-induced displacement, refugee studies, and disaster research have learned that involuntarily displaced peoples face many similar challenges. Although the places and peoples are geographically and culturally distant and the sociopolitical environments and causes of dislocation dissimilar, there emerge a number of common concerns and processes. Displaced people of all descriptions must cope with the consequent stresses and the need to adapt to new or radically changed environments. All may experience hardship, loss of homes, jobs, and the breakup of families and communities. All may suffer the endangerment of structures of meaning and identity. All must mobilize social and cultural resources in their efforts to reestablish viable social groups and communities and to restore adequate levels of material and cultural life.

We also identified the key components of resettlement planning as developed across a number of fields, assessing the utility of various approaches for crafting appropriate standards and strategies for future catastrophe-driven mass relocation. Central to these tasks are the issues of rights, poverty, vulnerability, and other forms of social marginality that are intrinsically linked to displacement. An important additional notion is that the displaced must be seen as active social agents with their own views on rights and entitlements, which have to be considered in any displacement and in the planning and implementation of resettlement projects.

SECTION IV

Planning Strategies and Skills: Response, Recovery, and Reconstruction

- Overview
  While a catastrophe will certainly exceed available response, recovery, and reconstruction capabilities, Section IV presents planning strategies and skills an emergency manager can employ to mitigate the effects of such an event. Issues that are discussed include the use of volunteers and unconventional sources of assistance, implications on mass care issues and an understanding of the political challenges associated with recovery and reconstruction.
- Objectives: *By the end of this section, the reader should be able to:*
  - Analyze planning issues relating to the management of mass care, including proper procedures for managing mass casualties and fatalities (e.g., display knowledge that mass graves/cremation are usually inappropriate and unnecessary).
  - Categorize a risk management system for catastrophe response.
  - Discuss examples of the importance of flexibility in catastrophe response.
  - Appraise strategic thinking with regard to catastrophe response, recovery, and reconstruction (e.g., where do we want to be and how do we get there under the circumstances of a catastrophe?).

- Define planning needs for management of voluntary responders (NGOs, PVOs, and spontaneous volunteers).
- Critique methods of integrating international responders into the response, recovery, and reconstruction effort.
- Compare the need to undertake mitigation and safety efforts before reconstruction work begins with the desire to speed the recovery.
- Assess the politics of recovery and reconstruction, providing examples from post-Katrina in New Orleans and along the Alabama coast, as well as post-tsunami Indonesia.
- Appraise methods of developing stakeholder buy-in under conditions of social disarray.
- Critique various approaches of how the long-term recovery might be organized.

- Outline of Topics
  - Chapter 9: Response
    - Risk Management System for Catastrophe Response
    - Importance of Flexibility in Catastrophe Response
    - Strategic Thinking with Regard to Catastrophe Response
    - U.S. Catastrophic Response Assistance Framework
    - Planning Needs for Management of Voluntary Responders
    - Methods of Integrating International Responders into the Response Effort
    - Systemic Differences between Disasters and Catastrophes
    - Integration Strategies
    - Planning Issues Related to Managing of Mass Casualties
    - Planning Issues Related to Managing Mass Fatalities
    - Myths about Planning for Catastrophes
  - Chapter 10: Recovery and Reconstruction
    - Issues Faced Following a Catastrophe
    - Defining Catastrophe Recovery
    - Elements of Recovery
    - Catastrophic Recovery Process
    - Stakeholders and Their Roles in Recovery
    - U.S. Disaster/Catastrophic Recovery Assistance Framework
    - Rules and Understanding of Local Needs
    - Timing of Catastrophic Assistance
    - Horizontal and Vertical Integration
    - Catastrophe Recovery Planning
    - Catastrophic Recovery Plan
    - Risk Management System for Catastrophe Response
- Section IV Conclusion

CHAPTER 9

Response Planning

9.1 LEARNING OBJECTIVES

When completing this chapter, you should be able to:

- Analyze planning issues relating to the management of mass care, including proper procedures for managing mass casualties and fatalities (e.g., display knowledge that mass graves/cremation are usually inappropriate and unnecessary).
- Categorize a risk management system for catastrophe response.
- Discuss examples of the importance of flexibility in catastrophe response.
- Appraise strategic thinking with regard to catastrophe response (e.g., where do we want to be and how do we get there under the circumstances of a catastrophe).
- Define planning needs for management of voluntary responders (NGOs, PVOs, and spontaneous volunteers).
- Critique methods of integrating international responders into the response effort.

9.2 KEY TERMS AND PHRASES

- Disaster Mortuary Operational Response Teams (DMORTs)
- Geographic Information System (GIS)
- HAZUS-MH (HAZards United States-Multihazards)
- Interagency Coordinating Council on Emergency Preparedness and Individuals with Disabilities
- Local catastrophic planning committee
- Mass casualties
- Management by objectives

- National Organization on Disability's Emergency Preparedness Initiative
- National Logistics Staging Area (NLSA)
- Objective-based response management
- Risk assessment
- Special needs population
- Unified command structure
- Unified Coordination Group (UCG)
- U.S. Catastrophic Response Assistance Framework
- Volunteer management planning

9.3 RISK MANAGEMENT SYSTEM FOR CATASTROPHE RESPONSE

BOX 9.1 QUOTE

When we speak of risk acceptability, we are *always* talking about issues of political value and moral choice (Clarke, 1999, p. 101).

9.3.1 Allocation of Scarce Resources

Especially since Katrina, medical professionals in the United States have been concerned about the issues related to scarce resource allocation. To help address these concerns, a preevent planning approach is being attempted in New York. In this model, a task force has written a set of ethical and clinical guidelines and has released the guidelines to the public for comment (Powell et al., 2008). Other critical supplies also may be in short supply. For example, medications, surgical supplies, and even personal protective equipment may require controlled distribution. Who will make the difficult allocation decisions?

9.3.2 Risk Assessment

One method of allocation is to take a risk-based approach. The first step in such an approach is to conduct a risk assessment. Risk assessment is the process of describing and characterizing the nature and magnitude of a particular risk and includes gathering, assembling, and analyzing information on the risk. Risk assessment is a foundation of risk management and risk communication. In order to effectively manage risks and to communicate risks to the public, a clear understanding of the nature and magnitude of the risk at relevant exposure levels is necessary.

Risk assessment should employ the best available scientific and/or technical data. Credible science is characterized by: (1) objective analysis of data, including suitability of experimental design, appropriate uses of statistical tests, and careful attention to the uncertainties in data collection and interpretation; (2) appropriate consideration of underlying assumptions and limitations of theories and models used in the analysis and interpretation of data; and (3) peer review and publication in reputable scientific journals. However, it should be recognized that credible scientific studies may lead to honest differences in data interpretation and support of competing theories as well. Calculations based on the different theories may lead to risk estimates that are significantly different.

Posner (2004) describes the types of risks we face:

- Natural catastrophes (e.g., pandemic, asteroids, major floods, earthquakes, and hurricanes)
- Scientific accidents (e.g., particle accelerator event, nanotechnology event)
- Other unintended manmade disasters (e.g., global warming)
- Intentional catastrophes (e.g., acts of war, terrorism)

Risk associated with catastrophes can be put into two broad categories:

1. **Catastrophes where an underlying event impacts a defined geographic area.** Following the underlying event, the consequences, such as a mass migration, may worsen the impacts of the event. The underlying event for this type of catastrophe can occur:
   - Over short period of time, i.e., such as an earthquake, hurricane, or most floods
   - Over a long period of time, i.e., drought, climate change, sea level rise
2. **Catastrophes where an underlying event impacts an increasing geographic area.** This type of catastrophe typically involves a communicable disease where casual contact can result in transmission. Depending on type of incident and the areas affected, these are some of the diseases that may become widespread:
   - Pandemic influenza, cholera, typhoid, shigellosis, hepatitis A and E, dengue fever, malaria, typhus, leptospirosis, acute lower respiratory infection (ALRI), measles, meningitis, and tuberculosis.

FIGURE 9.1 HAZUS (HAZards United States) logo.

For different catastrophic risk profiles, different risk assessment tools have been developed. For assistance in assessing risks from diseases and human exposure to hazardous environs, you should work closely with local public health officials. For natural hazards, FEMA has developed a state-of-the art risk Geographic Information System (GIS)-based assessment methodology called HAZUS-MH (HAZards—U.S., Multihazard) (Figure 9.1). HAZUS-MH is a powerful risk assessment methodology for analyzing potential losses from floods, hurricane winds, and earthquakes. In HAZUS-MH, current scientific and engineering knowledge is coupled with the latest GIS technology to produce estimates of hazard-related damage before, or after, a disaster occurs.

Potential loss estimates analyzed in HAZUS-MH include:

- Physical damage: To residential and commercial buildings, schools, critical facilities, and infrastructure
- Economic loss: Including lost jobs, business interruptions, repair and reconstruction costs

Social impacts: Including estimates of shelter requirements, displaced households, and population exposed to scenario floods, earthquakes, and hurricanes There are a number of options for obtaining assistance with HAZUS. Many colleges and universities make use of HAZUS in their geography, engineering, and social sciences programs. The reader can contact local universities to see if they are willing to assist. HAZUS user groups have been established throughout the United States. The growing network of HAZUS user groups has proved to be a valuable mechanism for creating partnerships between public and private sector organizations and industry in which mutually beneficial alliances are formed, goals and objectives developed, and projects identified and executed. User groups throughout the country have found enormous benefits and opportunities to develop successful projects through collaboration, combining resources and the sharing of information, data, and tips on software usage. (For more information on HAZUS user groups, go to: http://www.fema.gov/plan/prevent/hazus/hz_reghugs.shtm.)

HAZUS-MH can be ordered, free-of-charge, from the FEMA Publication Warehouse. Download the HAZUS-M3 order form, fill it out, and send it to the warehouse. (To learn more about HAZUS, go to: http://www.fema.gov/plan/prevent/hazus/index.shtm.)

9.3.3 Risk Management

BOX 9.2 RISK MANAGEMENT QUOTES

- [Managers] "do not accept the idea that the risks they face are inherent in their situation. Rather, they believe that risks can be reduced by using skills to control the dangers" (Shapira, 1995, p. 73).
- "The experience of successful managers teaches them that the probabilities of life do not apply to them" (Shapira, 1995, p. 127).
- "The methods that had been employed successfully for the 243 previous major disaster declarations since January 2001 proved inadequate for Hurricane Katrina's magnitude" (Bush Administration White House Report, 2006, p. 50).

Some have observed that the 9/11 terrorist attacks, Hurricane Katrina, and the 2008–2009 financial collapse and other recent catastrophes are a wakeup call for the need to better prepare for natural and manmade disasters that are having wider effects as the world grows more interconnected. There is a concern among some experts in risk analysis that we don't have all the planning tools we need to assess and manage catastrophic risk. Even if experts could develop better statistical tools, there are human tendencies to be overcome, like the habit of downplaying some risks and overreacting to others. Reason becomes clouded when people assess emotion-laden events like terrorist attacks, making sound public-policy decisions on risk very difficult, according to participants in a recent Harvard University/Wharton School of Business conference titled, "The Irrational Economist." On December 4 and 5, 2008, 100 leading scholars—including Nobel Laureates—in the fields of decision sciences, economics of information, political economy, catastrophic risk management, and insurance gathered for this conference.

Using risk management techniques that worked in previous emergencies, or even in previous disasters, may prove futile or even detrimental in the environment of a catastrophe. Underlying assumptions and data may not be applicable to catastrophes, which tend to be intense, but have an infrequent recurrence interval.

Perhaps the single most important lesson for experienced emergency managers is that many aspects of a catastrophic event will be beyond their control. Even if an emergency management group had all the resources it needed, which is doubtful, to conduct a rigorous catastrophic planning effort, there are risk assessment and management constraints, such as the lack of large historical data sets on specific types of catastrophes and a lack of human experience with such events from which to base planning assumptions. This is especially true when trying to develop hazard vulnerability assessments for terrorism incidents.

This is not to say that all is lost. We do have some catastrophe data sets and the advent of high-speed computers and cloud computing is allowing modeling and simulation to move from the laboratory to practitioners. It is extremely important that catastrophic risk management planning include advanced planning for evacuation and postevacuation; and rapid decision making for evacuation implementation.

9.4 IMPORTANCE OF FLEXIBILITY IN CATASTROPHE RESPONSE

BOX 9.3 QUOTE

"Far too often, the process required numerous time-consuming approval signatures and data processing steps prior to any action, delaying the response. As a result, many agencies took action under their own independent authorities while also responding to mission assignments from the Federal Emergency Management Agency (FEMA), creating further process confusion and potential duplication of efforts" (Bush Administration White House Report, 2006, p. 52).

It isn't just the 2006 White House report on Hurricane Katrina that has been critical of the bureaucratic government processes that slowed response efforts. Other authors also have described how inflexibility was one of the factors contributing to the poor response to Hurricane Katrina (Harrald, 2006; Takeda and Helms, 2006).

Many organizational cultures require strict adherence to established protocols. The emergency management community is certainly not immune to this concern. When confronted by a catastrophe, these organizations will be forced to either adapt to a change in a decision-making culture or fail to effectively assist in the response, recovery, and reconstruction. Not only will organizations need to change their culture, successful response to catastrophes will require flexible organizational leadership. Experience shows that, if flexibility does not occur at the top, it may develop at the bottom of organization and lead to inefficient procedures that can lead to wasteful and untimely delivery of assistance. Catastrophic leadership issues are explored in greater depth in Section V of this book.

9.5 STRATEGIC THINKING WITH REGARD TO CATASTROPHE RESPONSE

BOX 9.4 QUOTE

"Federal, state, and local officials responded to Hurricane Katrina without a comprehensive understanding of the interdependencies of the critical infrastructure sectors in each geographic area and the potential national impact of their decisions. For example, an energy company arranged to have generators shipped to facilities where they were needed to restore the flow of oil to the entire mid-Atlantic United States. However, FEMA regional representatives diverted these generators to hospitals. While lifesaving efforts are always the first priority, there was no overall awareness of the competing important needs of the two requests" (Bush Administration White House Report, 2006, p. 61).

The need to be flexible in responding to a catastrophe can result in confusion and even chaos if not properly managed and executed. Strategic thinking is needed to overcome this potential. Box 9.5 presents critical success factors related to strategic thinking for various phases of a catastrophe (Bullen and Rockart, 1981; Harrald, 2006; Haddon, 1972).

9.5.1 Objective-Based Response Management

One of the most recognized methods of strategic thinking is management by objectives (MBO), which is a process of agreeing upon objectives within an organization by management and employees, who then understand what these objectives are within their organization. The term *management by objectives* was first popularized by Drucker (1954) in his book *The Practice of Management*. MBO is also one of the 14 proven management characteristics on which the Incident Command System (ICS) is based that contribute to the strength and efficiency of the overall system. (To learn more about how ICS makes use of MBO, go to: http://www.fema.gov/emergency/nims/ICSpopup.htm)

The essence of MBO is participative goal setting, choosing course of actions, and decision making. When used to measure employee performance, an important part of the MBO is the measurement and the comparison of the employee's actual performance with the standards set. Ideally, when employees themselves have been involved with the goal setting and choosing the course of action to be followed by them, they

BOX 9.5 CRITICAL SUCCESS FACTORS RELATED TO STRATEGIC THINKING

Phase	Community	Agency	Government
Preevent	• Sufficient public education • Community-level drills • Sufficient food and water for one week	• Mobilization and response plans are based on realistic scenarios • Mobilization capacity and capability are adequate to meet expected needs • Adequate resources are available for initial response in high threat areas • Interorganizational coordination is preplanned	• Domain awareness and detection capability are created and maintained • Mobilization capacity and capability are adequate to meet expected needs • Adequate resources are available for initial response in high threat areas • Interorganizational coordination is preplanned; stakeholders are identified
Event	• Local resources are rapidly and efficiently integrated into predetermined response organization • Evacuations proceed in orderly fashion	• Mobilized response resources are rapidly and efficiently integrated into predetermined response organization • Coordinated multiorganization, networked response system is established • Ability to manage the collection, synthesis, analysis, and internal and external distribution of information is established • Organizational and operational adaptability and agility are maintained	• Situational awareness is obtained and shared across distributed organizational network • Resources in place are capable of initial life and safety response • Resource mobilization is based on accurate estimate of need for people, funds, and equipment • Resource mobilization is governed by preplanned organizational structure and process

Phase	Community	Agency	Government
Postevent	• Repopulation occurs as resources and infrastructure become available	• Continuing needs are identified • Organizational learning is accomplished	• Continuing needs are identified • Plan for transition to local support of continuing needs is developed and followed • External resources are demobilized according to established plans and procedures • Resources are provided to support economic and social recovery • Organizational learning is accomplished

Source: Adapted from Bullen (2008) and Harrald (2006).

are more likely to fulfill their responsibilities. The principle behind MBO is to create empowered workers and volunteers who have clarity of the roles and responsibilities expected from them, understand their objectives to be achieved, and thus help in the achievement of organizational as well as personal goals.

MBO can be applied to catastrophic planning by first asking the following questions:

- Where do we want to end up? What is the desired end state?
- How do we get there under the circumstances of this particular catastrophe on this particular day?

What are our short-term catastrophic response goals?

- Survival
- Food and water
- Shelter
- Medical care

What are our longer-term catastrophic recovery and reconstruction goals?

- Socioeconomic issues
- Housing
- Medical care/health maintenance
- Creating/maintaining a sense of community

What are the variables that impact our ability to respond effectively to a catastrophe?

- Indigenous population
- Evacuees/refugees
- Other agencies
- Emergent organizations
- Continued presence of the threat

If you include these issues when using an MBO approach to catastrophic planning, you achieve flexibility without losing your sense of purpose and focus. To successfully apply MOB, consider the following principles listed in Box 9.6.

BOX 9.6 PRINCIPLES OF MANAGEMENT BY OBJECTIVES

- Specific objectives for each member
- Participative decision making
- Explicit time period
- Performance evaluation and provide feedback

(If you would like to explore further what defines real objectives, go to: http://www.iw3c2.org/WWW2004/docs/2p236.pdf)

9.6 U.S. CATASTROPHIC RESPONSE ASSISTANCE FRAMEWORK

As discussed in Section II of this book, the *Catastrophic Incident Annex* to the *National Response Framework* (NRF-CIA) establishes the context and overarching strategy for implementing and coordinating an accelerated, proactive national response to a catastrophic incident within the United States. Note that when we talk about state and local governments, this includes tribal and territorial government entities as well.

A more detailed and operationally specific *National Response Framework–Catastrophic Incident Supplement* (NRF-CIS) is published independently of the NRF and annexes. This document was developed as part of the now-defunct NRP, but is still in use as of the writing of this book. (For more information, see: http://cees.tamiu.edu/covertheborder/tools/NRP_CIS.pdf)

9.6.1 Scope of the NRF-CIA

A catastrophic incident, as defined by the NRF, is any natural or man-made incident, including terrorism, that results in extraordinary levels of mass casualties, damage, or disruption severely affecting the population, infrastructure, environment, economy, national morale, and/or government functions. A catastrophic incident could result in sustained nationwide impacts over a prolonged period of time; almost immediately exceeds resources normally available to state, local, and private sector authorities in the impacted area; and significantly interrupts governmental operations and emergency services to such an extent that national security could be threatened. These factors drive the urgency for coordinated national planning to ensure accelerated federal and/or national assistance.

Recognizing that federal and/or national resources are required to augment overwhelmed state and local response efforts, the NRF-CIA establishes protocols to preidentify and rapidly deploy key essential resources (e.g., medical teams, search and rescue teams, transportable shelters, medical and equipment caches, etc.) that are expected to be urgently needed/required to save lives and contain incidents.

Upon the occurrence of a catastrophic incident, or in advance if determined by the secretary of Homeland Security, the government will deploy federal resources, organized into incident-specific "packages," in accordance with the NRF-CIS and in coordination with the affected state and the incident command structure.

Where state or local governments are unable to establish or maintain an effective incident command structure due to catastrophic conditions, the federal government, at the direction of the secretary of Homeland Security, may establish a unified command structure, led by the Unified Coordination Group (UCG), to save lives, protect property, maintain operation of critical infrastructure/key resources (CI/KR), contain the event, and protect national security. The federal government shall transition to its role of coordinating and supporting the state and local government when they are capable of reestablishing their incident command.

The NRF-CIA is primarily designed to address no-notice or short-notice incidents of catastrophic magnitude, where the need for federal assistance is obvious and immediate, where anticipatory planning and resource prepositioning were precluded, and where the exact nature of needed resources and assets is not known. This, of course, begs the question: What is the framework for responding to a longer notice catastrophe? Appropriately tailored assets and responses identified in the NRF-CIS, as well as other select federal resources and assets, also may be deployed in support of a projected catastrophic event (e.g., a major hurricane) with advance warning in support of the anticipated requests of state and local governments.

9.6.2 Noteworthy Observations

A catastrophic incident will likely trigger a Presidential Major Disaster declaration and result in the secretary of Homeland Security or a designee implementing the NRF-CIA/CIS.

All deploying federal resources remain under the control of their respective federal department or agency during mobilization and deployment. Some federal departments and agencies have the authority, under their own statutes, to deploy directly to the incident scene. Federal resources arriving at a National Logistics Staging Area (NLSA) remain there until requested by state/local incident command authorities, when they are integrated into the response effort.

Federal assets unilaterally deployed to the NLSA in accordance with the NRF-CIS do not require a state cost share. However, in accordance with the Stafford Act, state requests for use of deployed federal assets may require cost sharing (note that this cost share can be waved by the president, which will probably occur in a catastrophe). For no-notice or short-notice catastrophic incidents, federal resources identified in the execution schedule of the NRF-CIS will be mobilized and deployed, unless it can be credibly established that an action listed is not needed at the catastrophic incident venue. If during a response, it is determined that the incident is catastrophic in nature, any remaining actions not originally initiated from the execution schedule will be initiated.

Under the NRF-CIS, states are encouraged to conduct planning in collaboration with the federal government for catastrophic incidents as part of their steady-state preparedness activities. The federal government, in collaboration with states and local governments, the private sector, and nongovernmental organizations (NGOs), continues developing proactive plans for activation and implementation of the NRF-CIA to include situations where the need exceeds or challenges the resources and/or capabilities of state and local governments to respond and where the federal government may temporarily assume roles typically performed by state and local governments.

The occurrence or threat of multiple or successive catastrophic incidents may significantly reduce the size, speed, and depth of the federal response. If deemed necessary or prudent, the federal government may reduce the allocation of finite resources when multiple venues are competing for the same resources. This is where difficult "life and death" decisions may well need to be made and where our ethics discussion in Chapter 3 and our upcoming leadership discussion in Chapter 11 would certainly both come into play.

The initial response to a catastrophic incident starts on a local level with the local and/or state responders. However, there may well be circumstances that exceed the capabilities of state or local authorities in which they are unable to initially establish or maintain a command structure for incident response. In these instances, accelerated federal response may be warranted, and DHS/FEMA will coordinate response activities until state and/or local authorities are capable or have reestablished their incident command structure.

9.6.3 Continuity of Operations (COOP)/ Continuity of Government (COG)

As we discussed in Section II of this book, following a catastrophic event, segments of state and local governments, as well as NGOs and the private sector, may be severely compromised. The federal government and its national partners must be prepared to fill potential gaps to ensure continuity of government and public and private sector operations. The incident may cause significant disruption of the impacted area's CI/KR, such as energy, transportation, telecommunications, law enforcement, and public health and healthcare systems. To overcome or at least mitigate their effects, all levels of government may well need to implement their COOP and COG plans in the face of a catastrophe. Similarly, private sector entities will most likely need to rely on their business continuity plans to maintain functionality.

9.6.4 Roles of NRF Emergency Support Functions (ESFs) in Catastrophic Response

Normal procedures for certain Emergency Support Functions (ESFs) under the NRF may be expedited or streamlined to address the magnitude of urgent requirements of a catastrophe. All ESFs must explore economies of scale to maximize utilization and efficiency of limited resources. In the case of a catastrophic incident, the federal government or other national entities are expected to provide expedited assistance in one or more of the following areas:

- **Mass Evacuations (ESF #5—Emergency Management):** While primarily a state and local responsibility, federal support may be required for large-scale evacuations, organized or self-directed, that may occur. There also may be a need for evacuation of large numbers of people, including patients in local hospitals, nursing homes, and extended care facilities, as well as those with special needs, household pets, and service animals, out of the impacted area to safe areas in other states. Significant transportation and shelter coordination and resources may be required. There is likely to be significant shortage of response and casualty and/or evacuee reception capabilities throughout the impacted area. FEMA will support state(s) in evacuating pets and animals in a declared Major Disaster to the extent possible. Incident response efforts by state and local governments, as well as federal agencies, frequently involve air operations and flights for evacuation (ESF #5), medical (ESF #8), search and rescue (ESF #9), and public safety and security (ESF #13). In all cases, all operations must be coordinated with the Department of Transportation (DOT)/Federal Aviation Administration (FAA), which manages the nation's airspace and air traffic, before, during, and after a catastrophic incident.
- **Mass Care, Housing, and Human Services (ESF #6—Mass Care, Emergency Assistance, Housing, and Human Services):** The ability to support the provision of temporary shelter, food, emergency first aid, and other essential life sustenance to people, household pets, and service animals in the affected area may be complicated by contaminated resources or facilities and the inability to quickly transport resources into the area.
- **Public Health and Medical Support (ESF #8):** There is a significant need for public health and medical support, including mental health services. Medical and mental health support is required at medical facilities, casualty evacuation, embarkation, debarkation, and reception points and shelters and other locations to support field operations. In addition, any contamination requirement increases the requirement for technical assistance/resources.
- **Medical Equipment and Supplies (ESF #8):** Shortages of available supplies of preventive and therapeutic pharmaceuticals and qualified medical personnel to administer available prophylaxis are likely. Timely distribution of prophylaxis may forestall additional illnesses and reduce the impact of disease.

- **Casualty Transportation (ESF #8):** Federal resources may be required to manage the injured, exposed victims, and deceased if their numbers are extremely high.
- **Responder/Victim Medical Needs and Decontamination (ESF #8—Public Health and Medical Services):** State and local officials retain primary responsibility for responder/victim, screening, and decontamination operations. ESF #8 can provide technical assistance regarding how they can expand their capability to meet their medical decontamination requirements and can assist with patient evacuation to suitable locations.
- **Search and Rescue (ESF #9—Search and Rescue):** Resources and personnel to perform operational and tactical activities (e.g., locating, extricating, and providing onsite medical treatment to victims trapped in collapsed structures) are limited. If search and rescue operations are required in areas of contamination, the limited availability of properly equipped personnel and resources supports and underscores the need for prompt federal response. FEMA will make available its Urban Search and Rescue teams as needed.
- **Environmental Assessment and Decontamination (ESF #10):** Incidents involving a chemical, biological, or radiological weapon of mass destruction (WMD) may create significant environmental contamination, resulting in the immediate need to generate information on environmental contamination levels to support emergency decision making to ensure both public and responder protection. In addition, environmental decontamination and cleanup needs for buildings, critical infrastructure, and other areas may overwhelm state and local capabilities.
- **Public Safety and Security (ESF #13—Public Safety and Security):** Federal resources may be required to augment state and local governments in protecting the public and securing the impacted area. Law enforcement and emergency management officials who normally respond to incidents may be among those affected and unable to perform their duties. This federal role needs to be carefully considered in light of the posse comitatus restrictions on use of federal troops.
- **Public Information (ESF #15—External Affairs):** When state and local public communications channels are overwhelmed during a catastrophic incident, the federal government must immediately provide resources to assist in delivering clear and coherent public information guidance and consistent messages to the affected areas.
- **Critical Infrastructure/Key Resources (CI/KR Support Annex to the NRP):** CI/KR include the assets, systems, networks, and

functions that are vital to the American way of life. A terrorist attack on CI/KR or other natural or manmade disaster could significantly disrupt the functioning of government and business alike, and produce cascading effects far beyond the physical location of the incident. The federal government facilitates expedited information sharing and analysis of impacts to CI/KR, prioritized recommendations, and processes to consider incident-related requests for assistance from CI/KR owners and operators.

9.7 PLANNING NEEDS FOR MANAGEMENT OF VOLUNTARY RESPONDERS

BOX 9.7 QUOTE

"Volunteers have an important role to play in strengthening the capacity of local communities to resist the effects of disaster. Information exists to facilitate increased citizen involvement in disaster mitigation, but has not been effectively communicated to help individuals and organizations identify and embrace appropriate volunteer opportunities" (Points of Light Foundation, nd).

In conducting catastrophic planning, there are numerous issues associated with the use of volunteer responders. First, it is important to understand that it is quite possible that thousands of volunteers (including first responders) may self-deploy to the scene(s) of a catastrophe. Without appropriate management, volunteers may impede rescues, disrupt crime scenes, and may themselves become victims. If proper planning is done, experienced volunteer managers and coordinators may be capable of providing needed leadership.

9.7.1 Preevent Volunteer Management Planning

Preevent planning and training influences the availability (and value) of volunteers. Citizen involvement initiatives may help educate and prepare people to assist during catastrophic events. Preevent training programs and meetings also may serve as ways to help establish and document the volunteer's credentials. Drills/exercises should be used as opportunities to test plans for providing food, water, and shelter to volunteers as well.

After the catastrophe, citizens who are not volunteering may place more burdens upon strained resources. Some might observe that impacted citizens, if they're not part of the solution, are part of the problem. Anecdotes from as far back as ancient Athens describe people in catastrophic circumstances who feel helpless and hopeless, then essentially decide that "if they might be dead tomorrow, they're going to be drunk today."

9.8 METHODS OF INTEGRATING INTERNATIONAL RESPONDERS INTO THE RESPONSE EFFORT

BOX 9.8 QUOTE

"LESSON LEARNED: The Department of State, in coordination with the Department of Homeland Security, should review and revise policies, plans, and procedures for the management of foreign disaster assistance. In addition, this review should clarify responsibilities and procedures for handling inquiries regarding affected foreign nationals" (Bush Administration White House Report, 2006, p. 63).

Before Hurricane Katrina, the federal government had made little effort to formally plan for international assistance in response to a domestic catastrophe. Some argued that the United States was the greatest superpower in the world and couldn't possibly need help from other nations. International catastrophic assistance planning was limited to the role of the United States in assisting other countries.

The United States has participated through the United Nations, World Bank, and other international entities in planning for and assisting other agencies in their response and recovery from disasters. We are so committed to helping others that we maintain the U.S. Agency for International Development (AID), but there is no similar effort to coordinate potential assistance from other countries in helping the United States respond to and recover from a catastrophe.

During the Hurricane Katrina response, many countries offered response personnel, logistical resources, and funding to assist the United States and its impacted citizens. Much of this assistance was not accepted and, in some cases, foreign government complained that no representative of the U.S. government ever responded to their offer of assistance.

An interesting dynamic was that some senior federal officials, members of Congress and congressional staff, and some in the media expressed embarrassment that the United States, the "mightiest country on the face of the Earth," had to "stoop so low" as to accept "handouts" from other countries. Adding to this embarrassment was that some countries, seen as enemies of the United States, offered assistance in a thinly veiled attempt to score media points on the international stage.

There are numerous inhibiting factors to accepting international aid. Many of these factors are beyond the ability of state and local emergency managers to address, so it is important to consider their impact when developing a catastrophic plan that includes international assistance.

BOX 9.9 INHIBITING FACTORS TO ACCEPTING INTERNATIONAL AID

- **Language Barriers:** The inability to communicate commands and other relevant information to foreign responders can result in chaos and otherwise ineffective execution of the response mission. At worst, it could put these responders in harm's way.
- **Credentials:** Without preevent credentialing of foreign responders, emergency managers are left with the difficult decision of accepting help from foreign responders of unknown qualifications and of delaying their deployment while they are credentialed in the postevent environment.
- **Resource Compatibility:** Foreign-provided resources may be incompatible with existing organic resources. Communications devices may well not be compatible. Fire and rescue equipment is often not compatible (such as fire hose connections). Such items as medicines may be packaged and labeled in a manner inconsistent with U.S. Food and Drug Administration requirements.
- **Supply Chain Challenges:** The challenges with moving resources across long distances and across international boundaries following a catastrophe are legendary. These challenges become greatly amplified if there is political instability in the affected area that impact the ability to safely cross international borders and/or the safety of aid workers. Even in a stable political situation, import/export licenses and customs issues can slow international cooperation to a trickle.
- **Lack of Knowledge of U.S. Incident Management System:** Foreign aid workers could not be familiar with ICS, NIMS, and the NRF, making their integration into ongoing operations difficult at best.

9.9 SYSTEMIC DIFFERENCES BETWEEN DISASTERS AND CATASTROPHES

BOX 9.10 QUOTES

"Catastrophic Emergency means any incident, regardless of location, that results in extraordinary levels of mass casualties, damage, or disruption severely affecting the U.S. population, infrastructure, environment, economy, or government functions" (Homeland Security Presidential Directive 20).

"A catastrophic incident, as defined by the NRF, is any natural or manmade incident, including terrorism, that results in extraordinary levels of mass casualties, damage, or disruption severely affecting the population, infrastructure, environment, economy, national morale, and/or government functions. A catastrophic incident could result in sustained national impacts over a prolonged period of time; almost immediately exceeds resources normally available to state, local, tribal, and private-sector authorities in the impacted area; and significantly interrupts governmental operations and emergency services to such an extent that national security could be threatened" (Homeland Security Presidential Directive 20).

There are a number of systemic differences between disasters and catastrophes in how we respond. One of the more obvious is that many catastrophes will occur over geographical areas vastly larger than the areas impacted by most disasters. The result of this is that response resources will be spread over this vastly larger area, presenting challenges in command and control, communication, transportation, storage, and distribution. One result of having to deal with a large geographic area is that the ability to provide food, water, and shelter for victims (and responders) may be delayed for long periods of time.

The incident may cause significant disruption of the area's critical infrastructure, including communication and transportation networks, further negatively impacting the ability to effectively respond. Another difference is that federal and mutual aid resources may be unavailable for an extended time. Infrastructure damage may delay their arrival and the ability to sustain outside resources once they are in theater, and outside resources may not be available if they have to respond to the impacts of the incident on their own communities.

A sufficient number of properly trained and licensed medical personnel may be unavailable for long periods of time in a catastrophe. Medical personnel that reside within the impacted area may not be available for a number of reasons including having been evacuated and being unable to safely return; be assisting their own families; be hurt or killed. Even if most of the organic medical capability is available, the expected number of casualties from most catastrophes will most likely overwhelm the organic capability within the impact area. The availability of outside assistance may be slow in coming for reasons described above.

9.10 INTEGRATION STRATEGIES

BOX 9.11 QUOTE

"The federal response should better integrate the contributions of volunteers and nongovernmental organizations into the broader national effort. This integration would be best achieved at the state and local levels, prior to future incidents. In particular, state and local governments must engage NGOs in the planning process, credential their personnel, and provide them the necessary resource support for their involvement in a joint response" (Bush Administration White House Report, 2006, p. 49).

Emergency management officials must recognize that other response agencies may have drastically different organizational cultures as well as different policies and procedures. Although integration of leaders of response entities is essential, it may be more efficient to have individuals from the same agency work together in physical locations instead of splitting them up to integrate them into other agencies. Sending "Points of Contacts" or "emissaries" can achieve the desired effect of ensuring good open lines of communication.

Community and state-level exercises are invaluable opportunities to have personnel from disparate agencies learn more about the capabilities of other personnel and agencies and to build personal contacts. Exercises are also valuable opportunities to both check credentials and to emphasize the importance of proper credentialing.

9.11 PLANNING ISSUES RELATED TO MANAGING OF MASS CASUALTIES

This discussion will focus on the effective approaches to managing mass casualties resulting from a catastrophe. As with other parts of this book, we will focus on issues that differentiate between how we manage emergencies and disasters from how we manage catastrophes. In this discussion, we assume the reader has learned how to plan for and manage emergencies and disasters in prerequisite reading, training, exercise, or from personal experiences.

Managing mass casualties can be an unpleasant discussion and may represent some of the more difficult subject matter we will deal with in this book. Make no mistake that mass casualty management is one of the most important functions carried out during the response to a catastrophe. Failure to effectively manage mass casualties will most likely result in the spread of diseases, resulting in even more casualties and may result in infliction of unnecessary emotional distress on an already emotionally distraught population and can even lead to social unrest. Such unrest could actually threaten the continued viability of society's remaining order and institutions. Clearly, the management of mass casualties is a "necessary evil" that must be dealt with head on.

9.11.1 Creation of a Local Catastrophic Planning Committee

One of the most important steps that state and local officials can take to plan for a catastrophe is to establish a planning committee that represents a broad cross section of stakeholders. Some communities have struggled with this question when developing committees for issues such as bioterrorism, pandemic, earthquake, and hurricane planning.

Often in disaster planning, committees will include multidisciplinary representation; sometimes there are representatives of neighboring communities. For catastrophe planning, the committees must be multidisciplinary and must include representatives of a much larger geographic area, including neighboring cities, counties, states, and, sometimes, neighboring countries in border areas.

Disciplines to include in a catastrophic planning committee should include, at minimum:

- Emergency management
- Public health
- Health and medical
  - EMS
  - Hospitals
- Public safety
  - Police
  - Fire/Rescue
- Transportation infrastructure
- Public/private transportation providers

- Finance
- Legal
- Local elected officials
- Geographic areas
  - Neighboring cities and counties
  - States that may be impacted if your area is evacuated
  - If near an international border, cross-border government officials
- Federal agencies
  - HHS
  - DHS
  - DoD
- Private Sector
  - Major employers
  - Local trade groups/Chambers of Commerce
  - Potential suppliers of response assets and resources

9.11.2 Casualties in Catastrophe Area

We know that hospitals closest to but outside the disaster zone are particularly impacted by the event (Einav et al., 2004), so we must plan for those hospitals to be greatly impacted by the event. The use of "shuttered" hospitals has been advocated as a backup system during surge events (Zane et al., 2008), but it seems unlikely that this could present a viable option in many U.S. communities. It is possible that many casualties may have to be simply left behind as the event unfolds and sheltered in place and cared for to the extent possible. As discussed in Section I of this book, this occurred to a limited extent in Hurricane Katrina in Louisiana.

9.11.3 Special Needs Population

All levels of government and NGOs have an affirmative responsibility to address the special needs of the disabled. FEMA has been working with other federal agencies and NGOs for many years to address this issue. This effort took on added urgency following Hurricane Katrina with the passage of the Post Katrina Emergency Management Reform Act that calls for a more active planning effort to manage persons with disabilities.

In 2004, the Interagency Coordinating Council on Emergency Preparedness and Individuals with Disabilities was established to ensure that the federal government appropriately supports safety and security for individuals with disabilities in disaster situations. The council is overseen by the DHS.

The purpose of the council is to:

- Consider, in its emergency preparedness planning, the unique needs of agency employees with disabilities and individuals with disabilities whom the agency serves.
- Encourage, including through the provision of technical assistance, consideration of the unique needs of employees and individuals with disabilities served by state and local governments, and private organizations and individuals in emergency preparedness planning.
- Facilitate cooperation among federal, state, and local governments and private organizations and individuals in the implementation of emergency preparedness plans as they relate to individuals with disabilities.

The council maintains a Web site at: http://www.disabilitypreparedness.gov/index.htm

The National Organization on Disability launched the Emergency Preparedness Initiative (EPI) in 2001 to ensure that emergency managers address disability concerns and that people with disabilities are included in all levels of emergency preparedness, planning, response, and recovery (Figure 9.2). Their Web site has an excellent discussion of this issue at: http://www.nod.org/index.cfm?fuseaction=Page.viewPage&pageId=1564.

FEMA maintains a Web site concerning the special needs population at: http://www.fema.gov/plan/prepare/specialplans.shtm.

FIGURE 9.2 Emergency Preparedness Initiative logo.

9.11.3.1 Healthcare Providers

A minimal number of healthcare providers will likely remain operational in the catastrophe zone and essentially all will have left in a relatively short time following the event. Catastrophic event planning should assume that healthcare providers will be evacuated with patients and have appropriate resources for both. Many providers will require training on evacuation triage so decisions about who to leave behind can be made with some forethought and guidelines. It is extremely important to incorporate the national disaster medical system (NDMS) and any emergency management assistance compacts (EMACs) health assets fully into this planning process.

9.11.4 Impact on Surrounding Areas

Stories from past epidemics include instances where townspeople set up armed guards to prevent potentially infected/contaminated outsiders from entering the town. Similar incidents may occur in future catastrophes, particularly those related to biological, nuclear, or radiological events, where evacuees are seen as presenting a threat to the indigenous residents. Even the most prepared jurisdictions may have difficulty accommodating the needs of large numbers of evacuees arriving within their jurisdiction.

9.11.4.1 Economic

We must expect that many evacuees will be essentially destitute. The less notice there is of the event, the greater the probability that evacuees will arrive with "just the shirts on their backs." While some may eventually be able to reclaim funds from bank accounts and other decentralized finance institutions, many may lose all of their wealth that was associated with real estate and local investment if the event causes significant damages to the built environment. Some 500,000 people required temporary housing assistance and associated support following Katrina; many required assistance for years after the event. As of this writing, Katrina evacuees continue to receive assistance from both government and NGO entities.

9.11.4.2 Health

Lack of medications is a particular threat for vulnerable populations. Following Hurricane Katrina, many evacuees were left without medications and without records of their medications. There were many stories of people going to new physicians and asking for a resupply of their "red" pills or their "blue" pills. The problem is compounded by some insurance policies that restrict prescriptions to a maximum of a 30-day supply. The stress associated with an evacuation will likely have health consequences even for otherwise healthy individuals. Here again, close coordination with NDMS during the planning process is essential as NDMS maintains large stockpiles of pharmaceuticals that are available for rapid deployment (Figure 9.3). In addition, the planning process should consider including contacts with major pharmaceutical companies and trade organizations that may be willing to donate medications following a catastrophe. One such trade organization is Pharmaceutical Research and Manufacturers

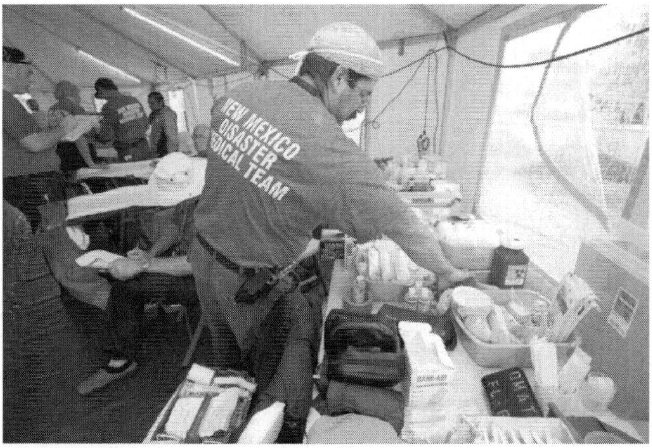

FIGURE 9.3 New Mexico Disaster Medical Assistance Team (DMATS) member retrieves medicine from medical supplies following Hurricane Frances in September 2004. (Photo by Andrea Booher/FEMA.)

of America (PhRMA) that represents the leading pharmaceutical research and biotechnology companies. (For more information on PhRMA, go to: http://www.rxresponse.org/Pages/default.aspx)

PhRMA is a partner in a broad healthcare industry disaster assistance program called *Rx Response* (Figure 9.4). Other partners include:

FIGURE 9.4 Rx Response logo. (Image courtesy of PhRMA.)

American Hospital Association
American Red Cross (ARC)
Biotechnology Industry Organization
Generic Pharmaceutical Association
Healthcare Distribution Management Association
National Association of Chain Drug Stores
National Community Pharmacists Association

Rx Response partners are committed to working together with federal, state, and local officials as well as volunteer organizations to help support the continued delivery of medicines to people who need them in the event of such an emergency, whether it is caused by a natural disaster, terrorist incident, or health emergency, such as a pandemic.

Rx Response partners include the drug and biotechnology manufacturing and distribution industries as well as hospitals and community pharmacies. *Rx Response* is a free public service that works with the ARC, Health and Human Services (HHS), and the DHS and state officials to share information to help support the continuing provision of medicines to patients during a severe public health emergency. To learn more about *Rx Response*, go to: http://www.rxresponse.org/Pages/default.aspx

9.11.5 Evacuation Issues

Following the South Asian tsunami, over 500,000 residents of the city of Banda Aceh in Indonesia were displaced; the pretsunami population was 4.8 million (WHO, 2008).

A stricken area may completely lose its infrastructure and government personnel as the catastrophe progresses. As evacuees flood neighboring cities, counties, and states, officials in these areas may essentially become local incident commanders. How do they coordinate? When multiple governors are involved, who is in charge? Even in the Washington D.C. multijurisdictional area, this question remains largely unresolved, years after the 9/11 attacks (Sheridan, 2008).

Response to and control of past events have been impacted by residents who refuse to obey official orders; in some cases, evacuating when ordered not to and, in other cases, not evacuating when the order was issued. This is an important issue for local Catastrophic Planning Committees to tackle.

9.12 PLANNING ISSUES RELATED TO MANAGING MASS FATALITIES

While used in response to catastrophes in some parts of the world, in developed nations, mass graves/cremation are usually inappropriate and unnecessary. Inappropriate or ineffective mass fatality management exacerbates stress and increases recovery time for survivors. Mass disposal of human remains is extremely traumatic to survivors and should be avoided at all costs. To further reduce trauma to surviving loved ones, religious considerations for victims and community should be honored to the maximum extent possible. Before human remains are disposed of, in cases where formal autopsies or burial preparation are not undertaken, photographic records and descriptions (e.g., gender, height, weight, etc.) of bodies and a list of personal effects found with the body, may help with later identification. Military graves registration best practices should be complied with to the maximum extent possible. (For more information, see http://www.dtic.mil/cgi- bin/GetTRDoc?AD=ADA428539&Location=U2&doc=GetTRDoc.pdf)

9.12.1 Mass Graves/Cremation Usually Inappropriate and Unnecessary

There is a misperception that human remains represent a health risk that must be abated by rapid disposal. The Word Health Organization in *Human remains do not pose a risk of communicable disease epidemics after natural disasters* (WHO, 2005) found that following the 2004 tsunami, Aceh Province, Indonesia, reported that over 200,000 people died (WHO, 2008). It took two months to bury the dead, yet no epidemics were reported and even among those workers specifically dedicated to caring for remains, there was no increased rate of disease (Morgan et al., 2006). There is no scientific basis for rushing to dispose of remains to the point where you disrespect religious wishes, fail to provide next-of-kin the opportunity to identify and collect the remains of the loved ones, fail to collect information that can be used to later identify remains that were identified at the time of burial/disposal, or otherwise treat remains in a manner that is traumatizing to loved ones.

One occupational risk did arise during the disposal of bodies in Thailand that was unexpected. Some workers that were disposing of human remains suffered from "heat stress and dehydration due to overuse of personal protective equipment, such as respirators" CDC, 2005). It seems strange that the only harm to response workers came from equipment intended to protect them from harm.

9.12.2 Federal Assistance for Mass Fatality Management: Disaster Mortuary Operational Response Teams (DMORT)

The NRF calls for the NDMS, as part of HHS, under ESF #8, Health and Medical Care, to provide victim identification and mortuary services. These responsibilities include:

- Temporary morgue facilities
- Victim identification
- Forensic dental pathology
- Forensic anthropology methods
- Processing, preparation, and disposition of remains

In order to accomplish this mission, DMORTs were developed and are composed of private citizens, each with a particular field of expertise, who are activated in the event of a disaster. DMORT personnel are required to maintain appropriate certifications and licensure within their discipline. When personnel are activated, all states recognize their licensure and certification, and the personnel are compensated for their duty time by the federal government as a temporary federal employee. During an emergency response, DMORTs work under the guidance of local authorities by providing technical assistance and personnel to identify and process the deceased.

Teams are composed of funeral directors, medical examiners, coroners, pathologists, forensic anthropologists, medical records technicians and transcribers, fingerprint specialists, forensic odontologists, dental assistants, x-ray technicians, mental health specialists, computer professionals, administrative support staff, and security and investigative personnel. It is important to point out that unlike DMATs that have a specific team makeup, DMORT configuration varies based on the needs of the incident (type of disaster/catastrophe). This can be problematic because a disaster medical assistance team (DMAT) is basically self-sufficient for up to three days while DMORTs rely on state and local governments to meet their resource requirements from their point of arrival on scene.

The DMORT program maintains three Disaster Portable Morgue Units (DPMUs). These are staged at locations on the East and West Coast for immediate deployment in support of DMORT operations. The DPMU is a depository of equipment and supplies for deployment to a disaster site. It contains a complete morgue with designated workstations for each processing element and prepackaged equipment and supplies. To learn more about DMORTs, go to: http://www.hhs.gov/aspr/opeo/ndms/teams/dmort.html.

9.13 MYTHS ABOUT PLANNING FOR CATASTROPHES

There are many myths about planning for catastrophes. To avoid making possible significant planning assumption errors, one would be well advised to avoid subscribing to these myths. At the conclusion of this book, you should be well versed in why these myths are not true and how to avoid them through effective planning and collaboration.

Myth: Catastrophes are just big disasters.

Reality: When hundreds of thousands or millions die, millions more are displaced (perhaps forever) and critical national infrastructures are broken for months or years or may never completely recover. The rules become completely different from disaster response. In some cases, impacted economies take a generation or more to recover.

Myth: I'm in charge.

Reality: Hurricane Katrina proved that even the president of the United States cannot personally "manage" a catastrophe. In the early days of the Katrina response, when the governor of Louisiana and the president failed to reach an agreement for a unified command, an effective response became impossible.

Remember, most believe that Katrina barely met the definition of a *catastrophe,* if it did at all. The only hope to effectively respond to a true catastrophe is through collaborative efforts of wide-ranging disciplines, agencies, communities, and leaders.

Myth: We're ready for anything that comes our way.

Reality: We'll never be fully ready for the broad spectrum of catastrophes that may befall us. However, we can become more prepared today than we were yesterday.

9.14 DISCUSSION QUESTIONS

1. Identify examples of how incidents, such as a pandemic or massive hurricane, might require a flexible response.
2. How might a current policy or procedure (such as ARC's policy of not opening shelters in floodplains) be inadequate or inappropriate for such a large-scale event?
3. In the wake of Katrina, offers of assistance poured in from across the globe. What policies and procedures would be needed to best integrate international assistance in future catastrophes in the United States?
4. How should we have handled the ethical dilemma of being embarrassed about accepting international aid, especially from countries that have been unfriendly toward us? What comes first, nationalistic pride or the needs of those being harmed by the catastrophe?
5. Identify who should be on the mass casualty planning committee in your community. Decide the disciplines to include as well as the geographic and agency representatives.

REFERENCES

Bullen, C.V., and J.F. Rockart. 1981. *A primer on critical success factors*. Cambridge, MA: MIT. Available at: http://dspace.mit.edu/bitstream/handle/1721.1/1988/SWP-1220-08368993-CISR-069.pdf?sequence=1. (Accessed January 10, 2010)

Bush Administration White House Report. 2006. The Federal Response to Hurricane Katrina: Lessons Learned, February. Washington, D.C.: GPO. Available at: http://www.whitehouse.gov/reports/katrina-lessons-learned.pdf. (Accessed January 10, 2010)

Centers for Disease Control and Prevention. 2005. Health concerns associated with disaster victim identification after a tsunami—Thailand, December 26, 2004–March 31, 2005. *Morbidity and Mortality Weekly Report* 54: 349–352.

Clarke, L.B. 1999. *Mission improbable: Using fantasy documents to tame disaster,* p. 101. Chicago: University of Chicago Press.

Department of Homeland Security. (nd) *Catastrophic incident annex.* Available at: http://www.learningservices.us/pdf/emergency/nrf/nrp_catastrophicincidentannex.pdf. (Accessed January 10, 2010)

Department of Homeland Security Presidential Directive 20: National Continuity Policy. (nd) Available at: http://www.dhs.gov/xabout/laws/gc_1219245380392.shtm.

Drucker, P.F. 1954. *The practice of management.* New York: HarperCollins.

Einav, S., Z. Feigenberg, C. Weissman, D. Zaichik, G. Caspi, D. Kotler, and H.R. Freund. 2004. Evacuation priorities in mass casualty terror-related events: Implications for contingency planning. *Annals of Surgery* 239(3): 304–10.

Haddon. W. Jr. 1972. A logical framework for categorizing highway safety phenomena and activity. *Journal of Trauma* 12: 197–207.

Harrald, J.R. 2006. Agility and discipline: Critical success factors for disaster response. *The Annals of the American Academy of Political and Social Science* 604(1): 256–272.

Morgan, O.W., P. Sribanditmongkol, C. Perera, Y. Sulasmi, D. Van Alphen, et al. 2006. Mass fatality management following the South Asian tsunami disaster: Case studies in Thailand, Indonesia, and Sri Lanka. *PLoS Medicine* 3(6): e195. Available at: http://medicine.plosjournals.org/perlserv/?request=get-document&doi=10.1371/journal.pmed.0030195&ct=1. (Accessed January 10, 2010)

Points of Light Foundation. (nd) *Preventing a disaster within the disaster: The effective use and management of unaffiliated volunteers.* Atlanta, GA: Points of Light Foundation & Volunteer Center National Network Pub. Available at: http://www.nationalserviceresources.org/library/items/M3248. (Accessed January 10, 2010)

Posner, R.A. 2004. *Catastrophe: Risk and response.* Oxford. New York: Oxford University Press.

Powell, T., K.C. Christ, and G.S. Birkhead. 2008. Allocation of ventilators in a public health disaster. *Disaster Med Public Health Prep.* 2(1): 20–26.

Shapira, Z. 1995. *Risk taking: A managerial perspective.* New York: Russell Sage Foundation Publications.

Sheridan, M.B. 2008. Guide to evacuate region reveals limitations. *The Washington Post,* February 4: B01.

Takeda, M.B. and M.M. Helms. 2006. Bureaucracy, meet catastrophe: Analysis of Hurricane Katrina relief efforts and their implications for emergency response governance. *International Journal of Public Sector Management* 19(4): 397–411.

World Health Organization (WHO). 2008. *Epidemic-prone disease surveillance and response after the tsunami in Aceh Province, Indonesia.* Geneva: WHO. Available at: http://www.who.int/wer/2005/wer8018.pdf. Accessed December 12, 2008.

World Health Organization (WHO). 2005. *Tsunami affected areas, 2005: Communicable disease risks and interventions.* Geneva: WHO. Available at: http://www.who.int/diseasecontrol_emergencies/guidelines/tsunami_risk_assessment.pdf. (Accessed January 10, 2010)

Zane, R.D., P. Biddinger, L. Ide, S. Phillips, D. Hurd, L. Buatti, H. Eustace, M. Amsden, D. Carr, and A. Hassol. 2008. Use of "shuttered" hospitals to expand surge capacity. *Prehospital Disaster Medicine* 23(2): 121–127.

CHAPTER 10

Recovery and Reconstruction Planning

10.1 LEARNING OBJECTIVES

When completing this chapter, the reader should be able to:

- Appraise strategic thinking with regard to catastrophe recovery and reconstruction (e.g., where do we want to be and how do we get there under the circumstances of a catastrophe).
- Critique methods of integrating international responders into recovery and reconstruction effort.
- Compare the need to undertake mitigation and safety efforts before reconstruction work begins with the desire to speed the recovery.
- Assess the politics of recovery and reconstruction, providing examples from post-Katrina New Orleans and the Alabama coast, as well as posttsunami Indonesia.
- Appraise methods of developing stakeholder buy in under conditions of social disarray.
- Critique various approaches of how the long-term recovery might be organized.

10.2 KEY TERMS AND PHRASES

- Business Continuity Plan
- Catastrophic Recovery Plan
- Economic recovery
- Emergency Support Function #14—Long-Term Community Recovery
- Emergent groups

- Environmental Planning and Historic Preservation Program
- Environmental recovery
- Horizontal and Vertical Integration Typology
- Model of recovery activity
- Physical recovery
- Social recovery
- Timing of disaster assistance
- U.S. disaster/catastrophic recovery assistance framework

10.3 OVERVIEW

The purpose of this discussion is to describe the disaster recovery and reconstruction process following a catastrophe. Emphasis is placed on defining the similarities and differences of disaster and catastrophe recovery processes following localized events. Stakeholders and their roles in recovery are addressed followed by an analysis of pre- and postevent recovery planning and policymaking. A proposed U.S. disaster recovery assistance framework is described as a means to integrate the topics previously discussed and stimulate critical thinking on the part of the reader. Note that when state and local governments are discussed in this chapter, territorial and tribal governments are included as well.

10.4 ISSUES FACED FOLLOWING A CATASTROPHE

In addition to describing the catastrophic recovery and reconstruction process, we will discuss how the process is not necessarily the same for all stakeholders involved. This discussion includes how the speed of recovery and reconstruction may vary according to differing levels of pre- and postevent planning, social vulnerability, and access to resources, including information. We also will explore differing recovery and reconstruction outcomes, including the degree to which hazard mitigation concepts are incorporated into the process and the factors facilitating sustainable catastrophe recovery and reconstruction (Smith and Wenger, 2006).

Recovery and reconstruction from a catastrophe are among the most complex aspects of hazards management and the least understood component of what are commonly referred to as the four phases of emergency management: preparedness, response, mitigation, and recovery. This holds true for both practitioners and hazards researchers. One of the greatest challenges is the lack of clarity as to who is actually in charge of long-term recovery and reconstruction activities following a disaster. The challenges associated with recovery and reconstruction are further exacerbated in the case of a catastrophic event.

It is important to note that virtually all of this discussion comes from research done on disasters, not catastrophes. This is a result of an exhaustive literature search that has found little scholarly research and writings related to recovery and reconstruction following a catastrophe. The author has attempted to extrapolate from both scholarly writings and his own extensive disaster experience, where possible, and has noted where catastrophes will likely differ from the experience with disasters.

10.5 DEFINING CATASTROPHE RECOVERY

Disaster recovery can be defined as: "The differential process of restoring, rebuilding, and reshaping the physical, social, economic, and natural environment through preevent planning and postevent actions" (Smith and Wenger, 2006). While this definition is related to disasters, it is also a good fit for catastrophic recovery.

This definition highlights the fact that recovery involves more than just the physical reconstruction of damaged structures following a catastrophe. It also includes social, economic, and environmental elements. *Catastrophe recovery* can be analyzed across the same dimensions (physical, social, economic, and environmental) that are used to describe noncatastrophic events. This does not mean, however, that unique factors do not come into play during these larger, more complex events.

Disasters can cause both positive and negative outcomes. Positive outcomes may include:

- The reassessment of past and projected development patterns in known hazard areas and taking steps to limit the impact of future events. This is typically referred to as *hazard mitigation*. The Federal Emergency Management Agency (FEMA) defines hazard mitigation as a "sustained action taken to reduce or eliminate the long-term risk to human life and property from natural disasters and their effects" (1995).
- Educating local residents, business owners, and community officials about the vulnerability they face in their community.
- Taking advantage of the postdisaster "window of opportunity" to enact new policies and programs addressing identified problems and vulnerabilities. This may include the adoption of more rigorous building codes, the construction of improved affordable housing, and/or the conversion of hazard-prone neighborhoods to open space.
- Increased investment in the area.
- Decreased levels of social conflict, as a catastrophe may serve as a unifying event.

Having listed these possible positive outcomes of a disaster, it is doubtful that some or many of these positive outcomes would pertain to a catastrophe, given the high severity level and broad range and length of suffering resulting from catastrophes. Negative outcomes may include:

- The degradation of environmentally sensitive areas.
- The failure to adequately assist socially vulnerable populations, resulting in their emigration or worsened economic and psychological condition.
- The long-term or permanent closure of local businesses.
- Reduced investment in areas prone to recurring disaster.
- Increased levels of social conflict among those competing for scarce resources found in the postcatastrophic environment.
- Collapse of settlements in some catastrophe-struck regions.

10.6 ELEMENTS OF RECOVERY

Physical recovery refers to the repair and reconstruction of damaged housing, public facilities (i.e., schools, police and fire stations, recreational facilities, etc.), infrastructure (i.e., water and wastewater treatment facilities, water and sewer lines, electric lines, roads, bridges, etc.), and businesses (including small operations and corporations). The act of engaging in physical recovery is often referred to as *reconstruction*. During reconstruction, the physical and spatial elements of human settlement patterns may be changed or "reshaped" based on decisions made by public officials, individuals, developers, and investors following a catastrophe. This is one of the potential positive outcomes of a catastrophe.

Catastrophes are often defined by the magnitude of physical damages sustained in a given area. This may be measured by aggregated loss, or the percentage of damages sustained relative to the total area exposed to the hazard. In the event of a catastrophe, the level of physical damages may involve regional, state, national, multinational, or global impacts.

Understanding the physical recovery of a community, state, or larger region (in the case of a catastrophe) requires recognizing the interconnectivity of housing, public facilities, infrastructure (or lifelines), and businesses. For example, in order to move back into one's place of residence, it must be deemed structurally sound and habitable by local building officials, which enables the servicing utilities (i.e., water, electric, and natural gas) to be turned back on. The repair of water and sewer systems allows schools and businesses to consider resuming operations, assuming necessary facility repairs have been made. Many residents with children will not return to their communities until schools are reopened, while the construction and engineering businesses normally responsible for making repairs to damaged structures and infrastructure must themselves be capable of resuming operations. This dependence on the reopening of schools before community repopulation takes hold was clearly seen in the repopulations patterns observed in the New Orleans area following Hurricane Katrina.

Social recovery includes the degree to which the restoration of social bonds and networks occur. Social networks often provide a key means of sharing information such as recovery grant and loan eligibility and other relief programs. Physical isolation and dislocation following a catastrophe often disrupt these networks. Catastrophes also can result in the formation of new informal ties as victims may work together to address a common problem. In some cases, this may result in the formation of what are referred to as *emergent groups.*

Catastrophes, unlike smaller events, can permanently alter social networks and result in the increase in large segments of an affected population's level of social vulnerability. Specific indicators of changes in social recovery may include the long-term or permanent scattering of large numbers of the population impacted by a catastrophe.

Economic recovery is closely tied to the ability of businesses to reopen and governmental activities to resume as they tend to comprise the principal economic drivers in most communities. Generally speaking, small, locally owned businesses are more vulnerable to the impacts of catastrophes than larger businesses and corporations who are able to spread their risk across many business locations. Several factors contribute to the higher level of vulnerability among small, locally owned business, including:

- The lack of suitable financial reserves.
- Inability to spread risk across multiple business locations.
- Inadequate insurance coverage.
- The failure to develop a preevent continuity of operations plan.

A Business Continuity Plan is comprised of a series of actions intended to help prepare a business to maintain operations or quickly recover in the event of a catastrophe. In the case of catastrophes, events may be of such magnitude and duration that large businesses, including corporations, may cease operations. Specific elements may include:

- A vulnerability assessment, which involves the estimation of the likelihood an event could impact their operations and others upon which they rely (vendors, suppliers, support infrastructure/lifelines, etc.). The vulnerability assessment includes a review of the structure in which the business is located and its supporting infrastructure relative to hazards and their forces (e.g., high winds, ground motion, flooding, fire).

- A list of proposed actions intended to reduce their exposure to damaging events based on this assessment. Action items may include reviewing the adequacy of existing insurance policies, the identification of an alternative business site, the purchase of a backup power source, the strengthening of the facility/structure in which the business is located, and the identification of alternate vendors and suppliers.

The ability of a community to recover economically is also inextricably linked to the restoration and functionality of supporting infrastructure. The physical repair of businesses represents one step in a more complicated process. The resumption of business operations requires the continuation or resumption of other businesses that supply needed goods and services upon which that business depends. It also requires the reopening of roads, other transportation networks, and communication channels to deliver those goods and services.

Depending on the nature of the business affected, a catastrophe can provide a substantial increase in the demand for their services. Examples include the following:

- Construction industry.
- Sale of building materials.
- Engineering and architectural firms.
- Disaster/catastrophe-based contractors (who pick up or manage debris cleanup efforts, write and administer disaster recovery grants, and supplement federal, state, and local staff following disasters).
- Others may see a dramatic short-term reduction in demand, followed by a sharp increase. These include automobile sales to replace damaged vehicles and retailers who sell items needed to refurbish homes (furniture, in-home appliances, etc.).

In other cases, many businesses suffer significant short-term, mid-term, long-term, and, in some cases, permanent economic hardships as demand for their products and services that are associated with the expenditure of disposable income may be severely curtailed. Examples could include the sale of luxury items and the tourism industry.

A catastrophe may impact national or even the global economy depending on the event's scope and duration. Examples may include impacts to a nation's gross domestic product or significant impacts on national or global markets that can take years to a generation to fully abate.

Environmental recovery, as understood in the context of a catastrophe, can be assessed relative to its impact on the environmental systems' ability to resume their principal preevent functions. The scope (geographic scale), duration, and severity of a catastrophe can affect larger environmental units, such as watersheds, ecosystems, and cross multiple political boundaries, which each may use differing means to address the event. Catastrophic environmental damages may include the following examples:

- The long-term or permanent degradation or loss of wetlands in coastal areas or floodplain. *Note*: wetlands serve a number of important functions; they are a natural buffer from the impacts of hurricanes and flooding, filter pollutants, and are an important habitat/nursery for aquatic life that serves as a food source for people and other animals.
- The long-term or permanent loss of a given area for agricultural purposes due to the impacts of a disaster. Examples may include desertification, long-term drought, climate change, or environmental contamination due to hazardous waste or nuclear radiation.

10.7 CATASTROPHIC RECOVERY PROCESS

The catastrophe recovery process can be understood in a number of ways. Key issues to discuss include:

- The catastrophe recovery process emerges from the initial response to a catastrophe, which overlaps the early phase of recovery efforts.
- As described earlier in the chapter, the catastrophe recovery process involves more than the steps associated with physical reconstruction efforts.
- Following a catastrophic event, the recovery process can take more than a decade.
- The transition from short-term recovery to long-term recovery and reconstruction is often difficult, as it is less certain who is responsible for these activities.

Disaster recovery has been described by a number of hazard scholars as following a series of orderly, sequential phases, each one comprised of a set of clear activities. Among the most widely recognized description of the process is the model developed by Haas et al. (1977). The fact that a significantly improved model of the recovery process has not been developed since the late 1970s highlights the fact that scholarly research in this area remains limited. We have taken the license to extrapolate the key issues above to the catastrophe recovery process.

The model (Figure 10.1) is comprised of four periods: Emergency, Restoration, Reconstruction I, and Reconstruction II.

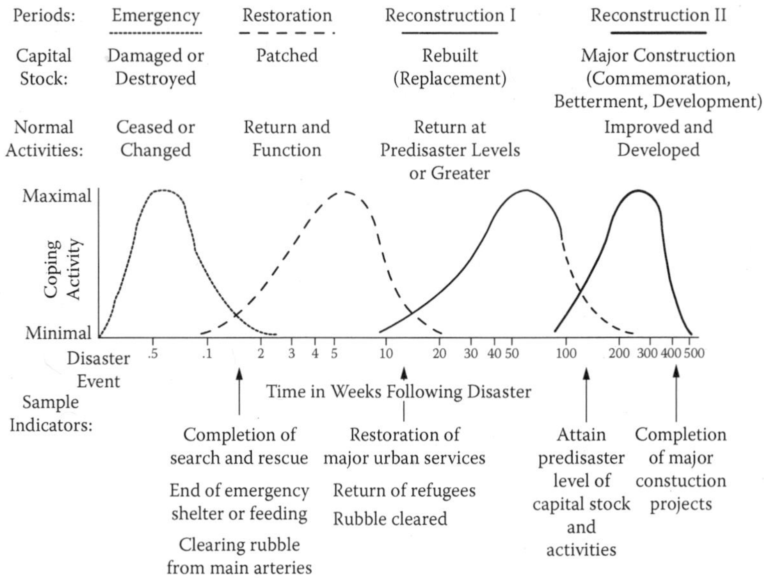

FIGURE 10.1 A model of the disaster recovery process. (From Haas, J.E., R.W. Kates, and M.J. Bowden (eds.). 1977. *Reconstruction following disaster.* Cambridge, MA: MIT Press. With permission.)

The *Emergency* period is comprised of search and rescue activities and mass feeding and sheltering of disaster victims. It also may include the clearing of disaster-generated debris from roadways.

The *Restoration* period may include the completion of the previously mentioned activities as well as the restoration of primary utility services (electricity, water, sewer), the return of evacuees, and the removal of most event-generated debris. A second phase of debris is typically generated as damaged and destroyed structures are demolished and taken to designated landfills and recycling centers.

The *Reconstruction I* period involves the repair and replacement of damaged structures and infrastructure to their preevent condition or greater.

The *Reconstruction II* period includes the completion of major construction projects and assumes that the impacted area achieves an improved result relative to preevent conditions.

The recovery timeline for a catastrophic event is of much longer duration. For example, the emergency/long-term temporary housing of disaster victims was extended for several years following Hurricane Katrina, which experts have debated as to whether or not to classify it as a catastrophe.

Other general critiques of the model include its failure to:

- Address the reality that catastrophes affect communities, groups, and individuals differentially because of varied levels of preevent planning/preparedness and social vulnerability. As a result, some may recover differently and to a lesser degree than others.
- The recovery process is not easily demarcated over time. Several factors facilitate or hinder the temporal aspects of recovery (and will be discussed in section 10.9, Disaster/Catastrophe Recovery Assistance Framework).
- The recovery process does not always follow a linear progression, rather it is more accurately characterized as a continuum based on the timing of assistance (as noted in section 10.9, Disaster/Catastrophe Recovery Framework), which is not uniformly provided.
- The model does not address preevent conditions, including variations in community capability to confront the challenges associated with recovery, nor does it consider the role of disaster recovery planning.

The definition posed by Smith and Wenger (2006) highlights the importance of pre- and postevent planning, which is discussed below.

10.8 STAKEHOLDERS AND THEIR ROLES IN RECOVERY

Stakeholder groups involved in disaster recovery are categorized accordingly:

- Public sector (federal, state, and local governments)
- Quasigovernmental and nongovernmental organizations (regional planning organizations, professional associations, colleges and universities)
- Nonprofit relief organizations (nonprofits and foundations)
- International aid organizations and nations
- Private sector and for-profit organizations (businesses and corporations, financial and lending institutions, insurance, media)
- Emergent groups and individuals

Each group is described briefly below, including their roles in recovery:

10.8.1 Public Sector

The *Public Sector* is comprised of federal, state, and local government agencies. Key federal agencies include, but are not limited to:

Department of Agriculture
FEMA/Department of Homeland Security (DHS)
Small Business Administration
U.S. Army Corps of Engineers
Department of Commerce
Department of Housing and Urban Development

To explore more about stakeholders and their roles in recovery, see the FEMA Emergency Management Institute Higher Education Program course, Holistic Disaster Recovery: Creating a More Sustainable Future. The course is available online at: http://training.fema.gov/emiweb/edu/completecourse.asp.

Congress also plays an important role in recovery following a catastrophe, as the members tend to appropriate large sums of federal dollars in the form of emergency supplemental appropriations. These funds are generally intended to address needs not met by federal programs associated with the Stafford Act. Congress, through existing and new subcommittees, may assess recovery activities and suggest changes in federal law or in the agencies tasked with disaster response and recovery-related duties.

Key state agencies and organizations include:

State emergency management agency
Governor's office
State legislature
Department of natural resources
Department of commerce or economic development
State budget office
State planning office
Department of public health

10.8.1.1 State Emergency Management Agency

Each state maintains an agency responsible for coordinating preparedness, response, mitigation, and recovery activities of the state before and after emergencies and disasters. Primary recovery roles include:

- Overseeing a comprehensive emergency management program.
- Developing and implementing a training and exercise program targeting state agency personnel, local government staff, community groups, and citizens. Areas of emphasis include preparedness, response, mitigation, and recovery-related issues.
- Coordination of state assets following a disaster.

- Direct liaison to FEMA pre- and postevent, often serving as an intermediary between local governments and FEMA following a disaster.
- Administrator of federal recovery programs following disasters.

State Emergency Management Agency recovery programs tend to focus on the means to administer federal recovery grant programs rather than a more proactive approach that includes a robust recovery-planning element.

10.8.1.2 Governor's Office

We previously discussed the role of the governor in some depth in Section II of this book, so this is just a quick recap. Under most state statutes, the governor is provided broad "emergency powers." This provision typically requires governors to adopt and implement an emergency management program focused on preparedness, response, mitigation, and recovery-related activities. These powers are substantial following a disaster or a catastrophe, providing significant discretion regarding the use of state assets and temporary modification of preevent rules and regulations (e.g., relaxation of environmental regulations associated with the burning of debris, relaxing weight limits on trucks hauling debris, relief supplies or construction materials).

Key governor roles associated with recovery include:

- Committing state assets (e.g., mobilizing the National Guard, other state agencies, and their physical assets)
- Declaring a state disaster
- Establishing evacuation routes
- Requesting federal assistance (through a request for a federal disaster declaration)
- Designating a state coordinating officer (typically the director of the State Emergency Management Agency)
- Entering into mutual aid agreements with other states
- Serving as the "public face" of the state—attempting to assure people and ensure that the state response and recovery efforts run smoothly
- Seeking additional postdisaster funding assistance through the state legislature and U.S. Congress

10.8.1.3 State Legislature

Key roles of the state legislature include:

- Approving requests from the governor to appropriate state disaster recovery assistance funding. These funds may be used

to assist with the nonfederal match requirement associated with many federal relief programs or the development of state programs intended to address gaps in federal assistance. A growing number of states have established a "rainy day" fund for this purpose.
- Establishing postdisaster recovery commissions tasked with the evaluation of past state performance following disasters. Their findings may result in modified policies and additional funding targeting identified weaknesses.
- Creating and passing state budgets that include funding for emergency management programs.

10.8.1.4 Local Government

Key local government actors and key tasks are:

- City manager/chief executive (oversee recovery efforts, hire contractors, manage staff, advise mayor, report on status of recovery)
- Finance official (track postevent expenditures, grant awards, and loans)
- Public works director (assess damages to public works, oversee the restoration of services and reconstruction efforts)
- Planning director (implement predisaster recovery plan, write postevent plan, assist with grants management activities)
- Police chief (protect public safety and monitor public and private property)
- Fire chief (search and rescue, fire suppression, assist in conducting preliminary damage assessments)
- Building official (lead damage assessment teams, conduct substantial damage determinations and habitability assessments; sign off on building permits)
- Floodplain administrator (assess flood-related damages, assess the accuracy of flood insurance rate maps, consider proposing increased floodplain management standards, notify individuals about techniques and/or requirements that may or must be incorporated into the repair or reconstruction of their home or business)
- Emergency manager (liaison with State Emergency Management Agency officials, coordinates the local requests for state assistance, particularly in the response and early recovery phase, may coordinate search and rescue activities. The local emergency manager also may be the fire chief or police chief). The role of the emergency manager in the recovery and reconstruction will vary widely from one community to another.

The local emergency manager often is most adept at preparedness and response activities, usually coming from a background in emergency medical services, fire, or police. This tends to result in a manager who is not particularly adept at dealing with recovery and reconstruction issues. The skills necessary to guide community rebuilding are often found in the building, zoning, environmental, land use regulatory, and the planning departments. The degree to which local land-use planners are involved in recovery varies widely across the country. In many cases, planners are not tasked with the development of a pre- or postevent recovery plan. Further hindering recovery planning efforts are the typically low levels of coordination between planners and emergency managers (Kartez and Faupel, 1994).

10.8.2 Quasigovernmental and Nongovernmental Organizations

Quasigovernmental and nongovernmental organizations (NGOs) include:

Regional planning organizations (such as Local Emergency Planning Committees)
Professional associations
Colleges and universities

10.8.2.1 Regional Planning Organizations

Regional planning organizations may be referred to as metropolitan planning organizations or councils of government and assume many roles that may be assumed by local governments. In many parts of the United States, regional planning organizations have become heavily involved in the coordination and writing of multijurisdictional hazard mitigation plans, but may lack legal authority to bind local jurisdiction to comply with their decisions.

Primary recovery roles include:

- Writing and implementation of local grant programs.
- Coordinating local land-use planning.
- Collecting and analyzing data associated with demographic, environmental, transportation, and economic development issues. Regional planning organizations are often tasked with the spatial analysis of this information.
- Assume local governance tasks.

10.8.2.2 Professional Associations

Professional associations provide the following services following a disaster/catastrophe:

- Critically analyze recovery activities and offer targeted solutions
- Provide expert opinion to other stakeholders involved in recovery
- Mobilize members to provide assistance based on their professional knowledge
- Conduct postdisaster damage assessments
- Provide pre- and postdisaster planning assistance
- Advocate for changes in building standards and codes

Professional associations involved in recovery include:

Association of State Floodplain Managers
National Emergency Management Association
International Association of Emergency Managers
International City/County Management Association
American Planning Association
American Institute of Architects

10.8.2.3 Colleges and Universities

Key roles in recovery include:

- Conducting hazards related research, such as HAZUS-MH (HAZard U.S.-Multihazards) analysis (that has been translated into practice)
- Teaching the growing number of students seeking a career in hazards and emergency management and homeland security
- Providing postevent technical assistance across a number of disciplines including planning, engineering, public health, and architecture

10.8.3 Nonprofit Relief Organizations

10.8.3.1 Nonprofits

Key roles of nonprofit organizations in recovery include:

- Provision of food, shelter/mass care, clothing, medical assistance, counseling, and crisis intervention
- Repair and reconstruction of damaged and destroyed homes
- Advocating for the protection of environmental systems
- Policy advocates
- Capacity building
- Technical experts

10.8.3.2 Foundations

Key roles of foundations in recovery include:

- Provision of gap funding and grants
- Sharing of information
- Mobilizing public opinion
- Identifying shortfalls and filling gaps in the recovery assistance system

10.8.3.3 International Aid Organizations and Foreign Nations

Key roles of international aid organizations and foreign nations include:

- Linking available international assistance with local needs
- Identifying appropriate diplomatic channels/organizations to funnel relief
- Drawing international attention to a catastrophic event

As we discussed in Chapter 9, the U.S. assistance network is not accustomed or designed to effectively receive aid from other countries. There is a greater chance such aid would be accepted during the recovery and reconstruction phase of a catastrophe as there is more time to move the aid through the complex diplomatic and bureaucratic process.

10.8.4 Private Sector and For-Profit Organizations

The Private Sector includes the following organizations:

Businesses and corporations
Financial and lending institutions
Insurance
Media

10.8.4.1 Business and Corporations

Key roles among businesses, corporations, and consultants include providing a number of important services, such as:

- Debris removal.
- Deployment of needed assets (food, water, and shelter).
- Writing and administering postdisaster assistance grant programs.
- Repairing and reconstructing damaged housing, infrastructure, businesses, and critical public facilities following a disaster.

- Local businesses may assist workers with temporary housing, child care, and the continued payment of salaries.
- Corporations may provide substantial relief donations in the form of money, supplies, or technical services associated with their line of work.
- Financial and lending institutions provide the majority of funding for postdisaster repair and reconstruction.
- Insurance-related businesses provide financial relief in the form of insurance-based settlements, which play a key role in financing reconstruction efforts.
- Media assumes the following roles:
  - Shape public perceptions
  - Influence policy formulation and modification
  - Inform disaster victims and other stakeholders involved in recovery

10.8.5 Emergent Groups and Individuals

Emergent groups and individuals assume the following roles:

- Share information, particularly with underrepresented or marginalized groups
- Advocate for the equitable distribution of postevent assistance
- Assert the nature of local needs
- Additional disaster/catastrophe recovery-related activities as identified by Stallings and Quarantelli (1985) include:
  - Limiting development in flood-prone areas
  - Replanting trees following a tornado
  - Opposing postevent housing approaches
  - Identifying funds to rebuild homes following a landslide
- Individuals may assume the following roles:
  - Share experiential lessons with others
  - Identify and articulate local needs
  - Serve as grass-roots activists demanding change to a system they see as flawed or inequitable

In catastrophes, some or many of the normally available governmental, nonprofit, and commercial organizations may not be present or able to function for some time in any given area. For this reason, emergent groups and international organizations may play a bigger role than is typically the case in disasters. This being the case, it would be wise for recovery planners to include contingencies for emergent and international groups.

10.9 U.S. DISASTER/CATASTROPHIC RECOVERY ASSISTANCE FRAMEWORK

Following disasters, a great deal of attention is placed on the assistance provided to individuals, groups, communities, and states, including the degree to which it is distributed based on:

- Race, class, and gender
- Power, influence, and political capital
- The impact of major events
- The differential access to information
- The definition of local needs, including those that have a temporal component
- How funding practices and policies contribute to an increased level of hazard vulnerability
- The important, but unrealized potential of planning for post-disaster recovery

Existing scholarly thinking has tended to lead to a narrow, disconnected understanding of the recovery process, limiting the broader applicability of observations and lessons learned, including the conditions that directly hinder the efficacy of existing pre- and postdisaster recovery planning efforts. The disaster/catastrophe assistance framework links key aspects of recovery, identifying members of the existing disaster assistance network, the types of assistance they provide and receive over time, and describes the critically important role of planning for recovery.

Recovery characteristics include the nature of the rules governing assistance, the timing of program delivery, the level of horizontal and vertical integration within and across organizations, and the varied understanding of local needs both before and after disasters/catastrophes. The framework includes three types of aid, including financial, policy-based, and technical assistance, and are provided by a fragmented network of differing stakeholder groups. The framework was originally developed by Gavin Smith, a previous director of the Mississippi Governor's Office of Recovery and Renewal following Hurricane Katrina to help explain the process of postdisaster assistance to staff. The Governor's Office of Recovery and Renewal focused on three primary goals:

1. Identify and procure financial assistance
2. Provide policy counsel to the governor, his cabinet, and local officials
3. Provide education, training, and outreach programs to state and local officials.

Members of the disaster assistance network include federal, state, and local governments; regional planning organizations; professional associations; the private sector; financial and lending institutions; the insurance industry; other nations; universities and colleges; the media; foundations and nonprofits; emergent groups; and individuals.

The practice of planning provides a process-oriented vehicle through which stakeholders can:

- Preidentify and access pre- and postdisaster funding, respectively
- Propose, modify, or create new policy
- Enact education, outreach, and training programs aimed at building local capacity and an enhanced level of self-reliance

The reluctance of stakeholders to effectively plan for recovery will be addressed in the context of the existing disaster/catastrophe assistance framework, providing insight into the conditions that help or hinder this process.

The material in this part of the discussion was largely derived from draft sections of *A Review of the United States Disaster Assistance Framework: Planning for Recovery* (Hubbard, 2008).

Members of the disaster/catastrophe assistance network include the following:

Public sector (federal, state territorial, tribal, and local governments)
Quasigovernmental and nongovernmental organizations (regional planning organizations, professional associations, colleges and universities)
Nonprofit relief organizations (nonprofits and foundations)
International aid organizations and foreign nations
Private sector and for-profit organizations (businesses and corporations, financial and lending institutions, insurance, media)
Emergent groups and individuals

Each member of the network may provide one or all of the following types of assistance:

- Funding
- Policy
- Technical assistance

The public sector of the disaster/catastrophe assistance network includes the federal government. In recent years, Emergency Support Function (ESF) #14 was created out of recognition that community recovery is an important aspect of the federal response and that recovery activities of federal agencies needs to be coordinated just as much as the response phase of disasters and catastrophes. There is myriad federal assistance available under many statutory authorities and the agencies that participate in ESF #14–*Long-Term Community Recovery,* and are committed to trying to coordinate the federal recovery assistance in an attempt to streamline the process for states and communities.

10.9.1 Emergency Support Function #14–Long-Term Community Recovery Annex

Emergency Support Function (ESF) #14–*Long-Term Community Recovery* provides a mechanism for coordinating federal support to state, regional, and local governments, nongovernmental organizations (NGOs), and the private sector to enable community recovery from the long-term consequences of large-scale disasters and catastrophes. ESF #14 attempts to accomplish this by identifying and facilitating availability and use of sources of recovery funding, and providing technical assistance (such as impact analyses) to promote community recovery and recovery planning.

ESF #14 may be activated for incidents that require a coordinated federal response to address significant long-term impacts (e.g., impacts on housing, government operations, agriculture, businesses, employment, community infrastructure, the environment, human health, and social services) to foster sustainable recovery. ESF #14 support will vary depending on the magnitude and type of event.

ESF #14 recognizes the primacy of affected state and local governments and the private sector in defining and addressing risk reduction and long term community recovery priorities, and in leading the community recovery planning process. ESF #14 long-term community recovery and recovery planning efforts will be coordinated with state- and local-level stakeholders. Federal agencies continue to provide recovery assistance under independent authorities to state and local governments, the private sector, and individuals, while coordinating assessments of need for additional assistance and identification and resolution of issues through ESF #14. It is important to note that ESF #14 excludes economic policymaking. The National Economic Council, the Council of Economic Advisors, and the Department of the Treasury develop all national economic stabilization policy.

Federal support is tailored based on the type, extent, and duration of the incident and long-term recovery period, and on the availability of federal resources. ESF #14 is not a funding entity, but facilitates the identification, coordination, and use of resources to support long-term recovery. The lead federal agency in the field is designated based on the type of event. Long-term community recovery efforts build resilience focusing on disaster resistance through permanent restoration of infrastructure, housing, agricultural industry, natural resources, community well-being, and the local economy, with attention to mitigation of future impacts of a similar nature.

ESF #14 provides the coordination mechanisms for the federal government to:

- Work with state and local governments, NGOs, and private-sector organizations to support long-term recovery planning for highly impacted communities.
- Convene interagency recovery expertise to provide strategic guidance to long-term recovery efforts.
- Identify and address long-term recovery issues, including those that fall between existing mandates of agencies.
- Avoid duplication of assistance, coordinate program application processes, and planning requirements to streamline assistance processes, and identify and coordinate resolution of policy and program issues.
- Identify programs and activities across the public, private, and nonprofit sectors that similarly support long-term recovery and promote coordination between them.
- Identify appropriate federal programs and agencies to support implementation of comprehensive long-term community planning and identify gaps in available resources.
- Identify appropriate federal programs and agencies to support and facilitate continuity of long-term recovery activities.
- Link recovery planning to sound risk reduction practices to encourage a more viable recovery.
- Strategically apply subject-matter expertise to help communities recover from disasters and catastrophes.

Under ESF #14, there are a number of primary federal agencies. When designated as the lead primary agency for a specific event, the lead primary agency is responsible for identifying areas of collaboration with support agencies and coordinates the integrated delivery of interagency assistance, issue resolution, and planning efforts. Each primary agency will lead planning efforts for areas of agency expertise and lead post-incident assistance efforts for areas of department/agency expertise.

BOX 10.1 PRIMARY FEDERAL AGENCIES

Agency	Functions
Department of Agriculture	Provides emergency loans and grants for the agricultural sector; economic and technical assistance for recovery of rural community facilities, businesses, utilities, and housing; technical assistance for agricultural market recovery, community planning, and community development; and resource conservation assistance
Department of Homeland Security	**FEMA:** Provides technical assistance in community, tribal, and State planning; recovery and mitigation grant and insurance programs; outreach, public education, and community involvement in recovery planning; building science expertise; and natural hazard vulnerability/risk assessment expertise **Office for Civil Rights and Civil Liberties:** Provides expertise in issues related to special needs populations to ensure that they are an integral part of the recovery process **Office of Infrastructure Protection:** Provides technical expertise in protective measures for critical infrastructure **Office of the Private Sector:** Provides expertise in private sector capabilities and services; provides coordination with private sector organizations **Transportation Security Administration:** Coordinates security of the Nation's transportation system in times of national emergency
Department of Housing and Urban Development	Provides building technology technical assistance, and assistance for housing, community redevelopment and economic recovery, public services, infrastructure, mortgage financing, and public housing repair and reconstruction
Small Business Administration	• Provides long-term loan assistance to homeowners, renters, businesses of all sizes, and nonprofit organizations for repair, replacement, mitigation, relocation, or code-required upgrades of incident-damaged property • Provides loan assistance to small businesses to address adverse economic impact due to the incident

10.10 FEMA'S ENVIRONMENTAL PLANNING AND HISTORIC PRESERVATION (EHP) PROGRAM

An important aspect of federal recovery and reconstruction assistance is environment planning and historic preservation. It is FEMA's policy to act with care to ensure that its disaster response and recovery, mitigation, and preparedness responsibilities are carried out in a manner that is consistent with all federal environmental and historic preservation policies and laws. FEMA uses all practical means and measures to protect, restore, and enhance the quality of the environment, to avoid or minimize adverse impacts to the environment, and to attain the objectives of:

- Achieving use of the environment without degradation or undesirable and unintended consequences.
- Preserving historic, cultural and natural aspects of national heritage and maintaining, wherever possible, an environment that supports diversity and variety of individual choice.
- Achieving a balance between resource use and development within the sustained carrying capacity of the ecosystem involved.
- Enhancing the quality of renewable resources and working toward the maximum attainable recycling of depletable resources.

The EHP program integrates the protection and enhancement of environmental, historic, and cultural resources into FEMA's mission, programs, and activities; ensures that FEMA's activities and programs related to disaster response and recovery, hazard mitigation, and emergency preparedness comply with federal environmental and historic preservation laws and executive orders; and provides environmental and historic preservation technical assistance to FEMA staff, local, state tribal, territorial, and Federal partners, and disaster assistance grantees and subgrantees. (To learn more, see: http://www.fema.gov/plan/ehp/index.shtm)

10.11 RULES AND UNDERSTANDING OF LOCAL NEEDS

An important defining characteristic of the disaster/catastrophe recovery assistance framework includes the nature of the rules guiding assistance and the level of understanding of local needs before and after disasters and catastrophes. Generally speaking, there is a poor connection between the rules governing aid distribution and local needs both before and following a catastrophe and to a greater extent following a catastrophe.

Some observations worth noting:

- Those who possess highly prescriptive rules-based relief programs tend to have a limited understanding of local needs.
- Those with the greatest understanding of local needs often operate fewer rules regulating their aid programs.
- There exists a significant "zone of uncertainty" among several organizations:
    - Our understanding of the types of assistance they provide is less understood (when compared to others).
    - The assistance provided does not neatly fall within the two defining parameters of the model. That is, the zone of uncertainly represents a greater degree of variability between the availability of resources and an understanding of local needs.
    - For example, a growing body of research shows that nonprofits and the private sector, including corporations and consultants, do, in fact, play a more important role in recovery than previously believed.
- Generally speaking, government entities have the greatest financial resources, while those at the local grass-roots entities have fewer financial resources.
- Rules tend to be shaped by a number of factors:
    - The level of communication among stakeholders.
    - The ability to share information by collecting, analyzing, and displaying data, the presence of politically powerful and influential advocates, and the use of a process-oriented forum to identify problems, share ideas, and propose solutions.
- The failure to create flexible programs among some members of the disaster/catastrophic assistance network is due, in part, to the current system that limits the sharing of information through preevent venues like recovery committees and participatory planning processes.
- Berke and Beatley (1997) describe an effective planning process as one in which those receiving assistance, the design of aid programs, and the capacity of organizations responsible for delivering it are well coordinated.

10.12 TIMING OF CATASTROPHIC ASSISTANCE

As demonstrated by Haas, Kates, and Bowden (1977) and Rubin and Barbee (1985), the primary research associated with the timing of disaster recovery has mostly focused on a description of the overall process and not on the timing of individual elements of the recovery process.

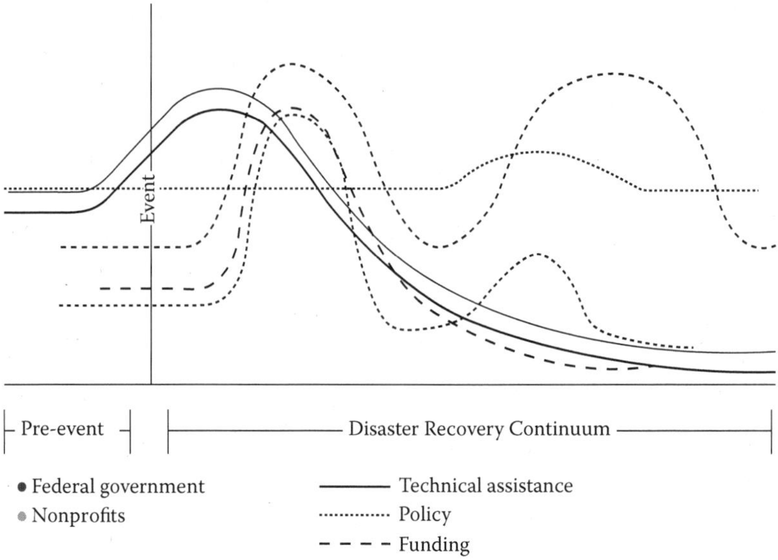

FIGURE 10.2 Timing of disaster assistance. (FEMA EMI Higher Education Program.)

Little emphasis has been placed on the timing of specific assistance programs provided by stakeholder groups over time.

Important aspects of the timing of assistance include:

- Differing members of the disaster/catastrophe assistance network provide disaster assistance (funding, technical assistance and policy implementation) at different times across the disaster recovery continuum.
- The timing of assistance significantly impacts disaster recovery outcomes.
- The inability to coordinate the timing of assistance can lead to the duplication of aid, the implementation of counterproductive programs, and missed opportunities to identify complimentary objectives.
- A key part of disaster recovery involves the appropriate balance between speed and deliberation (Olshansky, 2006).

Figure 10.2 represents a hypothetical description of the timing of assistance provided by two stakeholders in the disaster/catastrophe recovery process: the federal government and nonprofits. The differential timing of assistance has significant implications for community and individual-level recovery. For example:

- Following disasters, nonprofits often provide rapid assistance to homeowners (particularly the poor and other socially vulnerable populations).
- This assistance may take the form of housing repair and reconstruction.
- This assistance may occur before federal grant programs are provided to make similar repairs and hire the contractors necessary to perform the work.
- In many cases, the repairs may occur before local governments have decided whether to change building standards (i.e., increase elevation requirements in the floodplain, adopt more stringent building codes, etc.).
- As a result, well intentioned nonprofits may repair or rebuild damaged housing to their preevent condition.
- This may occur before federal grants to offset the costs of building to a higher standard are available, thereby missing the "window of opportunity" to incorporate hazard mitigation into recovery efforts.

10.13 HORIZONTAL AND VERTICAL INTEGRATION

The concept of horizontal and vertical integration provides a sound means to assess interorganizational relationships and their role in disaster recovery.

The horizontal and vertical integration framework describes the level of coordination across organizations and is particularly helpful in explaining a central aspect of recovery, that being the challenges associated with coordinating the range of programs provided by members of the disaster/catastrophe assistance network must be managed (May and Williams, 1986; Berke et al., 1993).

Horizontal integration involves close ties across organizations at the community level. Strong horizontal integration tends to influence the following:

- Active involvement of local members of the assistance network, including greater levels of public participation in decision making.
- Identification of pre- and postdisaster recovery needs.
- Creation of innovative solutions to disaster recovery issues and challenges.

Low levels of horizontal integration can lead to the following:

- Increased fragmentation among stakeholders, particularly those excluded from the decision-making process.

- The inability to create a sound, inclusive vision of what community recovery should be, often limiting the creation of innovative solutions and instead relying on narrowly defined federal assistance programs (in the case of an event that merits a federal major disaster declaration).

Vertical integration represents strong ties between local organizations and those located outside the community. Strong vertical integration can lead to the following:

- Better ability to influence policymaking. Relationships with state and federal agencies can shape and modify existing programs to better meet local needs.
- An important example includes the ability to influence the type of Congressional appropriations sought following major disasters.

Low levels of vertical integration can lead to the following:

- Limited influence over the policymaking and allocation decisions made by other organizations and agencies.
- The inability to effectively convey local needs to other organizations.

The vertical and horizontal typology includes four types (Figure 10.3). Each cell is indicative of the level of both horizontal and vertical integration.

		Vertical Integration	
			Weak
Horizontal Integration	Strong	Strong capacity and cohesion (self-organization) Type I	Cooperation Type II
	Weak	Type III Cooptation and coercion	Type IV Conflict resolution

FIGURE 10.3 Horizontal and vertical integration typology. (From Berke, P.R., J. Kartez, and D. Wenger. 1993. Recovery after disasters: Achieving sustainable development, mitigation, and equity. *Disasters* 17(2):93–109. With permission.)

Type I communities possess both strong horizontal and vertical integration, and:

- Are able to organize, recognizing the nature of external organizations and agencies that provide assistance following disasters.
- Are able to recover more quickly (all other factors being equal).
- Are able to effectively incorporate hazard mitigation into recovery.

Type II communities are characterized by strong horizontal and weak vertical integration. A good example of this type may include a tight knit rural community having limited interaction with external organizations, such as state and federal agencies. Type II communities:

- Are unaware of existing programs, their eligibility, and means to access them. As a result they are often heavily dependent on state and federal agencies for guidance and assistance.
- May possess strong local groups that provide assistance prior to a disaster, but are initially unaware of the existing channels of communication to external aid organizations.

Type III communities are characterized by weak horizontal and strong vertical integration, and:

- Have a good understanding of external assistance provided by others.
- May not employ locally inclusive, participatory decision-making strategies.
- Often rely heavily on external aid programs to the detriment of local involvement.

Type IV communities are characterized by weak horizontal and vertical integration, and they:

- Are less likely to be able to seek outside assistance or coordinate the actions of local organizations.
- Are less able to clearly identify local needs among varied stakeholder groups.

10.14 CATASTROPHE RECOVERY PLANNING

The benefits of disaster and, by extension, catastrophe recovery planning have been highlighted by a number of case studies (Schwab et al., 1998). For example:

- Berke et al. (1993): Interorganizational coordination (horizontal and vertical integration) is achieved through planning.
- Oliver-Smith (1990): Local needs are met through the use of inclusive planning techniques.
- Olson et al. (1998); Rubin and Barbee (1985): Incorporation of hazard mitigation into recovery.

It is worth noting that the benefits of planning for postdisaster recovery have been largely limited to qualitative case studies rather than quantitative approaches. One possible reason is that the benefits occur over many years, even generations, and researchers have been unable to collect sufficient quantitative data to assess the quantitative effects of recovery.

The disaster/catastrophe assistance framework highlights several problematic elements that affect recovery outcomes. Planning for recovery provides the means to address the prescriptive nature of rules governing assistance, the lack of understanding of local needs, the timing of assistance, and the level of horizontal and vertical integration. Practicing planners use a variety of tools and consensus-building techniques that are directly applicable to the act of pre- and postdisaster disaster and catastrophic recovery and reconstruction.

10.14.1 Predisaster

As listed in Box 10.2 (*Tools in the Redevelopment and Sustainability Toolbox)*, there are numerous tools (planning, zoning, subdivision controls, design controls, financial, and management) that are directly applicable to planning for pre- and postdisaster/catastrophic recovery. Employed before an event, many of the categories listed can mitigate the impacts of future events by serving to guide the type and location of development relative to hazards. Like in the case of disaster/catastrophic recovery, land-use planning tools and techniques are underutilized as a means to proactively reduce the impacts of hazards.

10.14.2 Postcatastrophe: Emergency Phase

In the immediate aftermath of a catastrophe, the land-use planner may be involved in the use of those tools and techniques listed in the "emergency" category. For example, a temporary building moratorium may be used to provide the time following a catastrophe to assess next steps, including the need to alter existing policies and past development patterns that continue to place people and property in harms way. A building moratorium also may serve to provide the time needed to develop a post-catastrophic recovery plan.

BOX 10.2 TOOLS IN THE REDEVELOPMENT AND SUSTAINABILITY TOOLBOX

Emergency

Damage assessment
Development moratorium
Temporary repair permits
Zoning for temporary housing
Prioritize infrastructure repairs
Trees and vegetation

Planning Tools

Acquisition
Easements
Infrastructure policy
Floodplain management plan
Environmental review
Annexation plans
Storm water management plan
Lending policies

Zoning Tools

Nonconforming issues
Performance standards
Special use permits
Historic preservation
Density controls
Floating zones
Overlay zones
Coastal Zone Management regulations
Floodplain zoning
Setbacks
Site plan reviews
Height and bulk regulations
Wetlands development regulations

Subdivision Controls

Subdivision regulations
Road width/access
Open space requirements

Design Controls

Design review
Building codes

Financial Tools

Targeting grant funds
Relocation aid
Special districts
Redevelopment projects
Transfer of Development Rights

Management Tools

Interjurisdictional coordination
Geographic Information System
Soil stability ratings
Public education

10.14.3 Postcatastrophe: Long-Term Recovery and Reconstruction

After a catastrophe, during long-term recovery and reconstruction activities, the planning categories are highly applicable. There are many tools that may be used in their current forms as part of a recovery plan, to impact reconstruction activities, or which can be amended based on public input or a reassessment of hazard vulnerability following the event.

Planners also emphasize the importance of the planning process and consensus-building techniques. These tools emphasize the importance of ongoing interaction, information collection and dissemination, and the search for mutually beneficial outcomes. Specific tools used may include:

Public participation techniques
Policy dialogue
Facilitation
Negotiated rule making

10.14.4 Precatastrophe

Seeking consensus and involving the public in decision making before a disaster can result in big dividends. Waiting until after the disaster to attempt informing the individuals (catastrophe victims) hinders the ability of stakeholders to engage in ongoing dialogue. That is to say, those who will be or are affected by proposed decisions, aid programs, etc., and those tasked with formulating policy should promote dialogue and communicate prior to a catastrophe.

The benefits of sound precatastrophe consensus building include:

- It allows for lengthy, continued discussions about issues that are important to diverse stakeholder groups.
- It allows for the identification of complimentary objectives ("win-win" solutions).
- It can prove to be an empowering experience for groups that have historically been disenfranchised or marginalized.
- It allows for the education of stakeholders about what to expect (and potentially change) following a catastrophe (e.g., complex, bureaucratic aid programs, the recovery process: search and rescue, debris removal, sheltering, temporary housing, building moratoria, damage assessments, housing, and infrastructure restoration and reconstruction, etc.).
- It allows for the creation of a collaboratively designed recovery plan that guides postevent actions.

10.14.5 Postcatastrophe: Emergency Phase

Immediately following a catastrophe, local governments and other members of the disaster assistance network are often overwhelmed with response-related activities. During this time, it is extremely difficult to get individuals to slow down enough to take the time to even think about planning for recovery.

Postcatastrophe (emergency phase) consensus-building issues and challenges include:

- If a recovery plan has not been developed before the event, it is incumbent on local officials to determine when it is appropriate to initiate postevent recovery and reconstruction planning activities.
- The best time to initiate varies based on the level of damage and access to common meeting places, materials, time, etc.
- Attempts to facilitate participation requires repeated and extensive efforts to reach out to all stakeholders impacted.
- Consensus-building techniques must reflect the fact that stakeholders may have suffered extreme hardship (loss of family, damaged or destroyed housing, loss of job, etc.).

In catastrophes, the emergency phase may be extended for quite some time and some of the normal participants in this process may not be available in many localities. Preplanning for the inclusion of emergent groups and for role flexibility among survivors may help minimize delays related to the catastrophe-caused absences and confusion.

10.14.6 Postcatastrophe: Long-Term Recovery and Reconstruction

Postcatastrophic planning for recovery and reconstruction is not the ideal time to discuss such far-reaching issues. However, it is better than the frequent alternative, which is to allow largely federal recovery programs to drive the local recovery options. Specific issues to consider when developing a postdisaster recovery plan include:

- Make sure to involve a wide range of stakeholders, including those from your community and external agencies and organizations providing assistance. Consider nontraditional partners (environmental and social justice groups, the private sector, the media, etc.).
- Recognize the catastrophe as an opportunity to address long-standing issues and problems in your community using multi-objective planning processes.
  - This may include taking a more aggressive posture on incorporating hazard mitigation into day-to-day actions and policies, addressing issues of equity/social vulnerability, etc.
- Take advantage of the postcatastrophe political context to state what the local needs are and what is lacking in available assistance.

- One idea that might be considered is the so-called aid auction or aid fair. The idea is that a central government (or, in some cases multilateral) organization calls together all the organizations, of all types, who are willing and interested in participating in a recovery project. Representatives of the government and citizens of the affected communities describe their needs and wants, and then the assembled groups "bid" (no money involved) on which part of what recovery process they want to participate in. Contracts are written that stipulate who is responsible for what. At the level of local reconstruction, ultimate control rests with the local government or its representative, but monthly coordination meetings are held for all "players" and concerned citizens. This has worked quite well in Nicaragua and even in a flood disaster in Poland (see: Krzysztof Chmura, et al.: *The Flood Aid Fair in Poland* that is available through the Urban Institute at www.urban.org/publications/409493.html).

10.14.7 Precatastrophe versus Postcatastrophe Planning

Planning for postcatastrophe recovery is best achieved prior to a catastrophe. This allows communities to identify existing vulnerabilities and develop strategies to address expected issues associated with community recovery and reconstruction activities. Attempts to develop recovery plans in the aftermath of a catastrophe, particularly those that attempt to alter the status quo are often met with significant opposition. Geipel (1991), in his study of the 1976 earthquake in northern Italy, discovered several significant findings:

- The disaster highlighted preevent social and economic characteristics, resulting in differing impacts based on this reality.
- Citizens sought to return to normal as quickly as possible.
- Citizens had a vision of what a postdisaster recovery plan should entail, which was recreating the city as it was before the disaster.
- The time associated with recovery and reconstruction was directly linked to preevent characteristics. For example, in those areas slated for preevent development, reconstruction activities reflected this reality. In poor communities, the economy further deteriorated following the earthquake.
- Planners had a limited *window of opportunity* to influence reconstruction outcomes.

10.15 CATASTROPHIC RECOVERY PLAN

Key components of a Catastrophic Recovery Plan include:

- An evaluation of existing local plans, policies, and programs and their relevance to recovery and reconstruction activities
- Legal authorities
- The creation of a Local Recovery Committee(s)
- Key topical elements commonly associated with a recovery plan include:
    - Damage and needs assessments
    - Postdisaster permitting
    - Building moratorium
    - Debris management
    - Restoration of public services
    - Repair of infrastructure
    - Critical facilities
    - Housing (emergency shelter, temporary and permanent)
    - Public health
    - Social services
    - Business/economic recovery
    - Hazard mitigation
- An implementation strategy that identifies those responsible for specific actions in the plan

In practice, pre- and postcatastrophe recovery planning is undertaken by a limited number of local communities. The reluctance to plan for recovery prior to a catastrophe results in several negative outcomes. They include:

- Poor coordination among stakeholder groups
- An increased length of time required for recovery
- Lower levels of public participation
- An increased likelihood that socially vulnerable populations will receive inadequate assistance
- A limited understanding of local needs among external aid organizations and agencies
- Low levels of vertical and horizontal integration
- Missed opportunities to incorporate hazard mitigation and other complimentary objectives into recovery and reconstruction
- An increased dependence on federal assistance following disasters

Polk County, Florida, has developed an excellent Postdisaster Redevelopment Plan. This plan has many aspects that can be translated to post catastrophic redevelopment planning. For more information, go to: http://www.polk-county.net/WorkArea/lin; http://www.polk-county.net/subpage.aspx?menu_id=226&nav=bus&id=9206kit.aspx?LinkIdentifier=id&ItemID=10068.

10.16 DISCUSSION QUESTIONS

1. An important question is who are the beneficiaries and losers following a catastrophe?
2. How do catastrophes impact the physical recovery or reconstruction of impacted communities?
3. Should Hurricane Katrina be categorized as a catastrophic event? Why or why not?
4. Are there other measures of physical, economic, social, and environmental impacts that should be considered? If so, what are they?
5. Does *Model of the Disaster Recovery Process* adequately address the issues associated with a catastrophic event? If not, what are important factors that are missing?
6. Is the *Disaster Recovery Assistance Framework* a good fit with catastrophic recovery processes and outcomes?

REFERENCES

Berke, P.R. and T. Beatley. 1997. After the hurricane: Linking recovery to sustainable development in the Caribbean. Baltimore, MD: Johns Hopkins University Press.

Berke, P.R., J. Kartez, and D. Wenger. 1993. Recovery after disasters: Achieving sustainable development, mitigation and equity. Disasters 17(2): 93–109.

Chmura, K., E. Strumillo-Kudiak, and M. Mikeelson. 1999. The flood aid fair in Poland: A Method to Promote Information Exchange. Washington, D.C.: The Urban Institute.

Federal Emergency Management Agency (FEMA). 1995. National mitigation strategy: Partnerships for building safer communities. Washington, D.C.: FEMA, p. 3.

Geipel, R. 1991. Long-term consequences of disasters: The reconstruction of Friuli, Italy, in its international context 1976–1988. New York: Springer-Verlag.

Haas, J.E., R.W. Kates, and M.J. Bowden (eds.). 1977. Reconstruction following disaster. Cambridge, MA: MIT Press.

Hubbard, J. (ed.). 2008. A review of the United States Disaster Assistance Framework: Planning for recovery. Emergency Management in Higher Education: Current Practices and Conversations. Washington, D.C.: Public Entity Risk Institute.

Hubbard J. (ed.). 2010. A review of the United States Disaster Assistance Framework: Planning for recovery (Forthcoming). Emergency Management in Higher Education: Current Practices and Conversations. Washington, D.C.: Public Entity Risk Institute.

Kartez, J. and C. Faupel. 1994. Comprehensive hazard management and the role of cooperation between local planning departments and emergency management offices. Unpublished paper, College Station, Texas A&M, Hazard Reduction and Recovery Center.

May, P. and W. Williams. 1986. Disaster policy implementation: Managing programs under shared governance. New York: Plenum Press.

National Response Framework. 2008. Federal Emergency Management Agency. Available at: http://www.fema.gov/emergency/nrf/ (accessed January 10, 2010).

Oliver-Smith, A. 1990. Post-disaster housing reconstruction and social inequality: A challenge to policy and practice. Disasters 14(1): 7–19.

Olshansky, R.B. 2006. Planning after Hurricane Katrina. Journal of the American Planning Association 72(2): 147–153.

Olson, R.S., R. Olson, and V. Gawronski. 1998. Night and day: Mitigation policymaking in Oakland, California before and after the Loma Prieta earthquake. International Journal of Mass Emergencies and Disasters 16(2): 145–179.

Rubin, C. and D. Barbee. 1985. Disaster recovery and hazard mitigation: bridging the intergovernmental gap. Public Administration Review 45(Special Issue): 57–63.

Schwab, J. ed. 1998. Planning for post disaster recovery and reconstruction. American Planning Association planning advisory service report 483/484.

Smith, G. 2008. A review of the United States disaster assistance framework: planning for recovery. In Emergency Management in Higher Education: Current Practices and Conversations, ed. J. Hubbard. Washington, D.C.: Public Entity Risk Institute.

Smith, G. and D. Wenger. 2006. Sustainable disaster recovery: Operational zing and existing agenda. In Handbook of disaster research, eds. H. Rodríguez, E.L. Quarantelli, and R.R. Dynes, pp. 234–257. New York: Springer.

Stallings, R.A. and E.L. Quarantelli. 1985. Emergent citizen groups and emergency management. Public Administration Review 45: 93–100.

IV Conclusion

While a catastrophe will certainly exceed available response, recovery, and reconstruction capabilities, Section IV presented planning strategies and skills an emergency manager can employ to mitigate the effects of such an event. Issues discussed included the use of volunteers and unconventional sources of assistance, implications on mass care issues and an understanding of the political challenges associated with recovery and reconstruction.

There are a number of systemic differences between disasters and catastrophes in how we respond. One of the more obvious is that many catastrophes will occur over geographical areas vastly larger that the areas impacted by most disasters. The result of this is that response resources will be spread over this vastly larger area, presenting challenges in command and control, communication, transportation, storage, and distribution. One result of having to deal with a large geographic area is that the ability to provide food, water, and shelter for victims (and responders) may be delayed for long periods of time.

In this discussion, we described the catastrophic recovery and reconstruction process following a catastrophe. In Section IV of this book, emphasis was placed on defining the similarities and differences of recovery processes following localized events, disasters, and catastrophes. Stakeholders and their roles in recovery were addressed followed by an analysis of pre- and postevent recovery planning and policymaking. A proposed U.S. catastrophe recovery assistance framework was described as a means to integrate the topics previously discussed and to stimulate critical thinking on the part of the reader.

In addition to describing the catastrophic recovery process, we discussed how the process is not necessarily the same for all stakeholders involved. This discussion included how the speed of recovery may vary according to differing levels of pre- and postevent planning, social vulnerability, and access to resources, including information. We also explored differing recovery outcomes, including the degree to which hazard mitigation concepts are incorporated into the process and the factors facilitating a sustainable catastrophe recovery.

Recovery from a catastrophe is among the most complex aspects of hazards management and the least understood component of what are commonly referred to as the four phases of emergency management:

preparedness, response, mitigation, and recovery. This holds true for both practitioners and hazards researchers. One of the greatest challenges is the lack of clarity as to who is actually in charge of long-term recovery activities following a catastrophe. The challenges associated with disaster recovery are further exacerbated in the case of a catastrophic event.

It is important to note that virtually all of this discussion came from research done on disasters, not catastrophes. This is a result of an exhaustive literature search that has found little scholarly research and writings related to recovery and rebuilding following a catastrophe. The author attempted to extrapolate from disaster experience where possible, and has noted where catastrophes will likely differ from disasters based on his extensive experience with disasters of all levels.

SECTION V

Essential Leadership Skills for Successful Catastrophe Management

- Overview
  If faced with a catastrophe, an emergency manager will need to utilize leadership skills that he/she may not have been called up before to use. Upon completion of Section V of this book, readers will be able to lead and influence others in a catastrophic situation by increasing their leadership skills and abilities.
- Objectives: By the end of this section, the reader should be able to:
  - Categorize communications styles that might be effective in a catastrophe.
  - Appraise problem-solving techniques that work in a catastrophe.
  - Analyze how effective political acumen helps an emergency manager be an effective leader in a catastrophe.
  - Correlate the need and techniques to effectively delegate responsibility during the lengthy catastrophic life cycle.
  - Appraise accountability mechanisms that will work in a catastrophe.
  - Contrast effective and ineffective techniques to maintain the required stamina to lead during the lengthy catastrophic life cycle.

- Critique individual differences in personal values and styles, and form generalizations about their impact on leadership behavior in a catastrophic situation.
- Assess your own style of exercising leadership and power as these styles relate to their roles in emergency management.
- Integrate knowledge about the different styles of leadership and influence and understand their impact on behavior in catastrophic context.

- Outline of Topics
  - Chapter 11: Leadership in a Catastrophe
    - Introduction
    - The Need for Skilled Crisis Leaders
    - Defining/Measuring Crises
    - Organizational Crisis Scenarios
    - Defining Leadership
    - Leadership's Role in a Crisis
    - Preparing for Crisis
    - Leading during a Crisis
    - Recovery and Rebuilding
    - What Leaders Can Do to Take Care of Themselves during a Crisis
- Section V Conclusion
- Epilogue

CHAPTER 11

Essential Leadership Skills for Successful Catastrophe Management

11.1 LEARNING OBJECTIVES

At the completion of this chapter, the reader should be able to:

- Categorize communications styles that might be effective in a catastrophe.
- Appraise problem-solving techniques that work in a catastrophe.
- Analyze how effective political acumen helps an emergency manager be an effective leader in a catastrophe.
- Correlate the need and techniques to effectively delegate responsibility during the lengthy catastrophic life cycle.
- Appraise accountability mechanisms that will work in a catastrophe.
- Contrast effective and ineffective techniques to maintain the required stamina to lead during the lengthy catastrophic life cycle.
- Critique individual differences in personal values and styles, and form generalizations about their impact on leadership behavior in a catastrophic situation.
- Assess your own style of exercising leadership and power as these styles relate to their roles in emergency management.
- Integrate knowledge about the different styles of leadership and influence and understand their impact on behavior in catastrophic context.

11.2 KEY TERMS AND PHRASES

- Center for Creative Leadership (CCL)
- Crisis Action Plan
- Crisis leadership
- Crisis lifecycle model
- Crisis management
- Emergency phase
- Emotional intelligence
- Hypercomplex crises
- Influence
- Preparation phase
- Rapid reflection forces
- Seven essential strategies
- Strategic planning cell
- Three "Cs" of effective crisis leadership: Communication, Clarity of vision and values, and Caring relationships
- Unconventional network of actors

11.3 OVERVIEW

If faced with a catastrophe, an emergency manager will need to utilize crisis leadership skills that he/she may not have had to use before. After reading Section IV of this book, the reader will know how to lead and influence others in a catastrophic situation by increasing his/her crisis leadership skills and abilities. Section IV will identify these essential skills and they will be applied to a catastrophic scenario case study: a major earthquake occurring within the New Madrid major earthquake fault zone.

An important element of the learning experience in this chapter is the analysis of various situations or cases related to leadership in a catastrophic incident. The scenario illustrates concepts that we have learned about in this book. Read the case study carefully, identify the main players or events in the case, determine the key issues or problems, and determine what actions should be taken to resolve or improve the situation.

11.4 INTRODUCTION

There has been much written about crisis management, but little has focused on crisis leadership. Most certainly, anyone that has had to face a catastrophe and to manage the response, recovery, and rebuilding would classify a *catastrophe* as a *crisis* based on any definition of the

FIGURE 11.1 May 29, 2009: President Obama demonstrated hands-on precrisis leadership by visiting FEMA headquarters to attend a meeting of the Homeland Security Council to mark the beginning of hurricane season. Federal agencies and departments briefed the president. (FEMA photo)

term. Managing a crisis and providing leadership in a crisis is not the same thing, although each addresses different aspects of a difficult situation. One might differentiate the two by saying that crisis management relates mainly to operational issues, while crisis leadership principally deals with how leaders handle the human responses to a crisis, including their own. We all have natural behavioral responses to crisis situations based on our needs and emotions. We may not be conscious of this, but our behaviors send messages to others about our underlying needs and emotions. It is within this set of behaviors that we find the core of crisis leadership (Figure 11.1).

For more than 15 years, the Center for Creative Leadership (CCL) has researched the key events that have shaped the careers and lives of executives. CCL is an international, nonprofit educational institution founded in 1970 to advance the understanding, practice, and development of leadership for the benefit of society worldwide. As a part of this mission, it publishes books and reports that aim to contribute to a general process of inquiry and understanding in which ideas related to leadership are raised, exchanged, and evaluated. *Note*: The ideas presented in its publications are those of the author or authors.

CCL's research indicates that hardships, such as those experienced during a crisis, can result in significant learning. In fact, 34% of the hundreds of managers CCL has interviewed indicated that their greatest learning occurred from hardships, which included leading in a crisis situation. CCL is the publisher of an important book by Klann entitled *Crisis Leadership* (2003) that is an excellent source on crisis leadership. Klann advocates using military lessons, organizational experiences, and the power of influence to lessen the impact of chaos on the people you lead.

Because people can learn from hardships, a crisis can develop personal and organizational leadership capacity by providing opportunities, such as:

- Hardships cause individuals and organizations to examine what is important, to further define or redefine their core values.
- Crises renew individuals and organizations by getting rid of the old unsustainable practices, thoughts, and policies, and bringing in the new.
- Crises bring out courage, honor, selflessness, loyalty, and many other positive behaviors.
- Individuals leading or otherwise involved in the crisis learn lessons about their own strength of character and how much adversity they can take.
- Handling a crisis promotes confidence and personal growth.
- What survives the crisis emerges better and stronger than before—it's tempered and hardened by the hardship.
- A crisis can create bonding and a keen sense of camaraderie and community among responders through the power of a shared experience.

For years, crisis management has been synonymous with reactive leadership. This stems from a belief that crises are both unpredictable and unexpected; however, Braden et al. (2005) argue in their Harvard University–Kennedy School of Government paper, Crisis: A Leadership Opportunity, that this is simply not true. Crises develop as an organization's values, beliefs, culture, or behavior becomes incongruent with its operating environment. A leader who is "tuned in" to the signals of impending crisis and understands how to harness the urgency brought on by the situation can minimize the potential dangers and maximize the resulting opportunities.

This chapter presents the Crisis Lifecycle Model as a generic representation of crisis. It illustrates that a crisis can be broken into three unique phases. In the first or *preparation phase,* the organization is typically mired in the comfort zone. Here, leaders struggle when introducing any change or learning, as the organization prefers to avoid conflict and sustain equilibrium. However, as a crisis hits, the organization is jolted into the *emergency phase,* often threatening its very existence. Once the immediate threat is eliminated, the organization enters the *adaptive phase*. In this phase, the leader has the attention and urgency to solve the underlying issues that caused the crisis in the first place. Unfortunately, many leaders don't take advantage of this opportunity and push the organization back toward the original status quo, leaving the organization exposed to experiencing the crisis again in the future.

The study of crisis leadership is becoming increasingly important as leaders in all walks of life face varying degrees of crises, spawning numerous recent books and articles. From this extensive body of work, Braden et al. and the other authors found seven essential strategies that are crucial for success. They include: (1) lead from the front, (2) focus on the core purpose, (3) build the team, (4) conduct continuous planning, (5) mitigate the threat, (6) tell the story, and (7) profit from the crisis. These strategies are very similar to those espoused by Klann (2003) in *Crisis Leadership*.

Lagadec, a Foreign Policy Institute Fellow at the School of Advanced International Studies (Center for Transatlantic Relations), at Johns Hopkins University, writes in *Unconventional Crises, Unconventional Responses: Reforming Leadership in the Age of Catastrophic Crises and Hypercomplexity* (2007) that there is a need to reform conventional crisis management.

Lagadec observes that since the 1990s North America and Europe have confronted a series of unconventional, catastrophic, or "hyper-complex" crises, e.g.;

North America

The 1998 ice storm in Canada
9/11
The anthrax crisis
The SARS outbreak in Toronto
The 2003 Northeast blackout
Hurricane Katrina

Europe

The "Mad Cow" disease
1999 storms in France
2002 floods in Central Europe
2003 heat wave
2007 forest fires in Portugal and Greece

In addition, both sides of the Atlantic responded to the 2004 Indian Ocean tsunami.

These events have shared striking similarities, especially inasmuch as they destabilized leaders in charge of response and reconstruction efforts. Lagadec (2007) believes that the time has come to launch a high-level, *balanced* dialogue among North American and European leaders, analysts and experts, representing the public, private, and humanitarian sectors, with the goal to enable a detailed and long-term exchange among their respective experiences, unresolved questions, intuitions, and proposals for reform.

According to Lagadec, recent catastrophic crises repeatedly have overwhelmed traditional mechanisms for crisis planning and management, and made them instantly obsolete, in several respects.

11.4.1 The Challenge of the "Unthinkable"

This series of events has clearly shown that complex Western societies today are not equipped to confront a major crisis effectively. The basic concepts and tools defined from the 1980s onward in the field of crisis management fall short of what is needed today. Klann (2003) argues the same, indicating that *crisis management* is not the same as *crisis leadership*. Current mechanisms for anticipation and response were designed to confront an event circumscribed to a specific area, within a global system that otherwise remains stable, but we lack a radar screen and a method to deal with crises that destabilize entire systems. This is *the* crucial challenge of our times. High intensity incidents today occur with unprecedented frequency—literally with respect to meteorology, but metaphorically in every other critical field as well. Moreover, because of structural characteristics of modern societies, e.g., strategic interdependencies, global connections, constant acceleration of causal chains, and the risk of "liquefaction" of our systems' foundations (i.e., the unthinkable collapse of trusted systems hit by unconventional events), massive destabilization can be provoked not only by catastrophic events, but also by crises that initially seem mundane, and would have been in the recent past.

11.4.2 The Culture of Leaders

Generally speaking, in all countries and sectors, they have proved culturally incapable of taking the "unthinkable" seriously, let alone reacting effectively when it actually occurs. They tend to eschew the challenges posed by unconventional events, in large part because those fall outside the intellectual principles that frame organizational architectures and "normal" decision making, and that underlie the selection of leaders in the first place. In this context, the very mention of unconventional risks and crises tends to provoke considerable uneasiness and reluctance. It is also extremely politically challenging to propose to use society's limited resources to plan for the "unthinkable" when we face so many challenges that are prima fascia, such as crime, unemployment, healthcare, etc. Cultural, analytical, and even psychological obstacles thus compound the challenge at stake.

11.4.3 The Identity of Leaders

The public sector's traditional monopoly on planning and response efforts time and again has shown its limits when confronted with unconventional events. The priority now must be to define new allocations of tasks and responsibilities among the public, private, and humanitarian sectors, as well as a wider public: a new "Social Contract," without which the democratic foundations of Western societies themselves are at risk, as Hurricane Katrina has shown in the United States.

11.4.4 Complex Maps of Actors

Catastrophic crises systematically involve an enormous variety of stakeholders, on an international scale (as, for instance, during the 2004 tsunami). These include spontaneous, unanticipated coalitions among unlikely partners on one end of the spectrum, as well as individuals that can wield extraordinary and unexpected power through their control of needed resources and especially through the channels of "old" and "new" media alike, at the other end. This complexity makes it nearly impossible for "traditional" leaders to plan, let alone coordinate response efforts.

11.4.5 New Processes for Crisis Recovery

Today's unconventional crises often do not contrast a single "Ground Zero" with an unscathed "outside" from which response can be safely organized; on the contrary, they often destabilize systems in their entirety. Therefore, instead of a clear succession of phases from planning to response to reconstruction (each under the leadership of a different agent, which withdraws and transitions to the next when its job is done), leaders now must tackle the three together, in other words, build reconstruction dynamics into their contingency plans, all the while taking into account that *leaders and responders themselves* might be among the victims of unconventional crises.

Lagadec advocates exploring paradigm shifts and cultural changes needed to confront chaotic and hypercomplex situations. Presenting the notion of Rapid Reflection Forces, i.e., teams of experts "parallel" to the incident command center (yet enjoying immediate access to its leaders), are tasked with taking a "step back" vis-à-vis the urgent requirements of crisis management, and adopting a different timeframe than the first responders, in order to think through strategic posture, anticipate unheeded obstacles, and advise leaders and first responders on these questions at each critical juncture in the response effort. You also might call this effort a "strategic planning cell," the key to which is to be thinking out ahead of the response/emergency (recovery) reconstruction life cycle of a catastrophe. In any given crisis, these experts must especially clarify:

- What the essence of the problem is.
- What the key traps to be wary of are.
- What unconventional network of actors needs to be set up.
- What critical initiatives will put the response effort in the best possible posture early on.

11.5 THE NEED FOR SKILLED CRISIS LEADERS

Nothing tests a leader like a crisis. There is an element of the leader's deepest character that is revealed during highly charged, dramatic events. A crisis can quickly expose a leader's hidden strengths and core weaknesses. It can show the world if the leader has what it takes to function effectively when the heat is on. Will the leader address the crisis head-on, take those actions needed to fix it, and, if appropriate, take responsibility for the crisis? Will the leader freeze, or worse, claim to be a victim and pass off the responsibility to others? What can and should a leader do to find out what went wrong and to ensure it doesn't happen again?

Klann (2003) identifies three key themes of crisis leadership and their impact on helping people and organizations through perilous times. These themes—communication, clarity of vision and values, and caring relationships—are certainly important to leaders in normal operations, but their importance is magnified during a crisis. By paying attention to these themes, leaders can hope to increase their understanding of practices that handle the human dimension of a crisis. The result is a leader more prepared to contain the crisis, regain control of the situation, ensure the minimum amount of damage is done to the organization, and effectively prevent, defuse, and reduce the duration of these extremely difficult leadership situations.

11.6 DEFINING/MEASURING CRISES

A crisis is generally characterized by a high degree of instability and carries the potential for extremely negative results that can endanger the continuity of the organization. It's a key moment or critical period that brings both surprise and dramatic change. In this way, a crisis can be described as a turning point in the affairs of an individual or an organization. It's significant because the consequences of the situation will be decisive in determining the future of that individual or organization. A crisis has the potential to divide an organization's past from its future, to replace security with insecurity, and to separate effective leaders from ineffective ones. A crisis also has the potential to swap routine for creativity and to shift an organization from "business as usual" into significant change.

Although no two crises are ever the same, they share some common traits. For example, a crisis isn't usually expected or planned for. It generally comes as a bombshell that frightens and stuns those on whom it falls. There may have been signs and indications of impending difficulties, but in the flow of daily operations, they were ignored, placed on the back burner, or wished away. The element of shock and even terror can be sharp and devastating if the crisis has an element of physical danger, if the crisis causes a death or serious injury, or if the crisis results in the destruction of property (for example, one's office or place of work). For these reasons, a crisis can exert a high impact on human needs, emotions, and behaviors.

In defining a crisis, it is worth noting that:

- A crisis can happen any time, anywhere, to any organization or community.
- Some are predictable, others come without warning.
- Most are characterized by a high degree of instability.
- Most carry potential for extremely negative results.
- A crisis brings about dramatic change.

11.7 ORGANIZATIONAL CRISIS SCENARIOS

Crises also have the tendency to bring a high degree of chaos and confusion into an organization. Typically, there is a lack of information precisely when virtually everyone in the organization has a huge emotional need for it. Those involved have a need to know and understand what happened, why it happened, and how it will impact their futures. Ambiguity is especially potent. High-stress situations, such as a crisis, can move usually rational people away from sense and reason. Common emotions arising in crisis situations include fear, anger, anxiety, sorrow, surprise, shock, disgust, love, and the desire for revenge.

These emotions can trigger positive or negative behaviors; the emotions themselves are not positive or negative, but the behaviors they trigger can be. People in a crisis can act with compassion, self-sacrifice, and courage, or they can display selfishness, cowardice, and greed. The potential for conflict and illogical behavior can be great. Previously dysfunctional behavior has the potential to become even more dysfunctional during a crisis. For those emotionally impacted by the crisis, even the simplest tasks can become difficult to perform. It is in this chaotic, ambiguous, and highly charged emotional environment, one rife with the human element, that leaders must lead and lead well.

11.8 DEFINING LEADERSHIP

Leaders must pay attention to the components of *influence*. Influence is the ability to persuade, convince, motivate, inspire, and judiciously use power to affect others in a positive way. Generally speaking, it's not the kind of authority that comes from leveraging title, position, or regulations. But exactly how is this different from other methods of leadership that managers carry out every single day? After all, the ability to influence others is an important part of leadership in both good and bad circumstances. The power of influence would seem to be a useful leadership skill no matter what the style of the individual leader (some managers are more inclusive than others and some are more autocratic, for example, in the way they approach their work). The difference lies not in the importance of influence as a leadership capacity, but rather in the particular context of the crisis itself, an emotional situation in which the components of influence can be defined in three key elements: communication, clarity of vision and values, and caring. Crisis leadership is a special case in which these specific tools of influence perform a critical role. In a crisis, timelines are more critical. There isn't as much time for reflection. Rapid decision making and a higher call to action become the norm.

In his 1989 book, *Leadership in Organizations*, Yukl lists several influencing tactics that people commonly use. Among these, he includes ingratiation, exchange (quid pro quo), coalition, inspirational and personal appeals, consultation, legitimizing, and pressure.

His description of these tactics suggests that managers can categorize them (and others can perceive them) as either positive or negative in their practice. For example, using pressure and micromanagement to achieve results (such as checking frequently on a direct report's progress with a specific assignment) can have a negative impact during the best of times and especially during a crisis. On the other hand, personal appeals based on a legitimate relationship between managers and direct reports (one constructed not just on rank and position, but on a common interest and vision) can also bring results and are more effective during a crisis.

All of these general influencing tactics are useful for a leader before, during, and after a crisis. Along with these tactics are personal influencing methods that leaders can practice and that also can be highly effective during a crisis. These personal methods can be grouped into skills, traits, and perspectives. Leaders can develop skills through training and through experiences like problem solving, decision making, and conflict resolution. Leaders can fine tune traits (individual characteristics) by paying attention to such areas as integrity, courage, and risk taking. Perspectives, which include the attitudes and points of view held toward leadership (if leaders take care of their direct reports, they don't have to worry about accomplishing the mission, for example), can be fuel for further reflection and learning in an effort to bolster leadership competence and confidence—things sorely needed in times of crisis.

Each skill, trait, and perspective is a useful tool for leading during a crisis. But they are even more effective when integrated into a single crisis leadership strategy. Consider how the following skills, traits, and perspectives might add to a leader's ability to get results through others even during times of crisis. Of the eight listed earlier in this section by Yukl (this list isn't exhaustive; many more skills, traits, and perspectives have an effect on the leader's ability to influence), the first three are most crucial to crisis leadership. Examples of the skills, traits, and perspectives that are drawn from civilian and military history illustrate how these capabilities form crisis leadership.

If one were to take one thing away from this discussion at this point, it is that emergency managers cannot rely on authoritarian tactics to get results during a catastrophe. This begs the question: Can an emergency manager effectively lead within the current U.S. emergency management framework in a true catastrophe? We explored earlier in this book as to whether the U.S. emergency management framework failed in the Hurricane Katrina response because it wasn't properly implemented or if it failed because it couldn't be implemented. The academic and practitioner emergency management community will continue to argue this matter for years to come. Nonetheless, it is worth taking a moment to compare NIMS/ICS principles to those of the crisis leadership (see Box 11.1).

BOX 11.1 COMPARING THE FEDERAL EMERGENCY MANAGEMENT FRAMEWORK WITH CRISIS LEADERSHIP

Federal EM Framework

- Command
- Control
- Coordination
- Highly structured; clear roles and responsibilities
- Training and exercises are conducted regularly

Crisis Leadership

- Leadership isn't just about title or position
- Leadership skills can be learned/honed
- Communication
- Clarifying vision and values
- Caring

11.8.1 Communication

A well-honed communication strategy is essential and critical to any organization before, during, and after a crisis situation. But there are also personal communications skills leaders can sharpen to make themselves more effective in those situations (Figure 11.2). Those skills include clear and articulate verbal expression that takes into account tone of voice, choice of words, and tempo of speech. Another is careful listening that involves appropriate eye contact, responsive gestures (such as saying "okay" or "sure"), not interrupting, and repeating key points to ensure understanding (known as "active listening"). Personal communication skills related to appropriate body language and a clear, concise, and straightforward writing style are also helpful during a crisis.

FIGURE 11.2 Winston Churchill during World War II boosted English morale during the London Blitz with his oratory and messages of hope along with a spirit of defiance against Nazism. (Photo from Imperial War Museum Collection.)

Leaders in a catastrophic crisis response must measure their words carefully. The wrong words will be examined and greatly amplified by the media. As Vince Covello pointed out in Box 11.2, *one negative statement is equal to three positive statements*. This is true because the media tends to amplify and rehash negative information more than positive information.

BOX 11.2 RISK COMMUNICATION

1 N = 3 P
(one negative statement is equal to three positive statements).

Vince Covello, PhD, Speaker
National Public Health Leadership Development Network
April 2003

As House Speaker Dennis Hastert learned (see Box 11.3), when people are struggling to recover from the devastating effects of a catastrophe, it is probably not the time to speak "ventured opinions" as to whether their homes will be rebuilt. Such opinions are better expressed as part of the dialogue concerning long-term reconstruction, which ensued months later, not within two weeks of the incident. Amplifying people's outrage was that Hastert was seen as an outsider who lacked the knowledge and qualifications to pass judgment on their future even though he was in a position of power over their future. In other words, he was the wrong person saying the wrong thing at the wrong time in the catastrophic life cycle. But yet, there was a positive effect, in that his comments galvanized the community's response with local leaders rallying to the city's defense (Box 11.3).

BOX 11.3 EDITORIAL: *THE TIMES-PICAYUNE* (SEPTEMBER 2, 2005)

"Even as people from New Orleans desperately search for their family members and rescue workers patrol the region in boats, hack through roofs and try to pluck survivors out, some people in other parts of the country have begun to blame us, the victims. Our crime? Choosing to live in New Orleans.

Especially heartless were U.S. House Speaker Dennis Hastert and the writers of an editorial that appeared Wednesday in the *Republican-American*, a newspaper in Waterbury, Conn. Mr. Hastert was quoted by the *Daily Herald* of Arlington Heights, Ill., saying it makes no sense to rebuild New Orleans where it is. "It looks like a lot of that place could be bulldozed," he said.

After Mr. Hastert made his insensitive comments, his press secretary tried to spin them. The speaker didn't mean that there shouldn't be a New Orleans, the spokesperson said. He was just suggesting that as they rebuild, officials give serious thought to how future destruction could be prevented. That goes without saying. We're much more sophisticated now than we were when the city was founded in the 18th century. Of course, our officials are going to rebuild in such a way that reduces the threat of future devastation.

At least President Bush realizes how valuable we are. He flew over the storm-ravaged areas of Louisiana, Mississippi, and Alabama on Wednesday afternoon and seems sincerely sorrowful for all the people whose lives have been irreversibly changed by this storm. His promise to send aid, and lots of it, was encouraging. It's going to take a huge amount of money to rebuild New Orleans and a similarly large amount of assistance to sustain the hundreds of thousands of people who have been displaced."

FIGURE 11.3 Martin Luther King, Jr. delivering his "I have a dream" speech at the Lincoln Memorial on August 28, 1963. (Photo National Archieves)

11.8.2 Clarity of Vision and Values

Having a clear vision and value system (either personal or passed down from the organization) that can be communicated so members of the organization understand it, feel ownership of it, and endorse it is a powerful influencing tool, before, during, and after a crisis (Figure 11.3). But clarity of vision is only effective if it is associated with a set of values that clarify what is important to the organization and what isn't. During a crisis, the leader can leverage a credible vision and value system and use both as a rallying point and as a way to provide stability to workers who are rocked by incidents.

11.8.3 Caring

A sincere interest and genuine concern for others goes a long way toward meeting the emotional needs of people experiencing a crisis. Just consider what it is like to be on the receiving end of an inconsiderate or uncaring leader. When normal, emotionally healthy people are treated with respect, dignity, approval, appreciation, attention, significance, value, and trust, they will generally respond in kind. During a crisis, the value of caring is greatly amplified as workers look to see if their leaders are compassionate (Figure 11.4).

FIGURE 11.4 In the wake of the 9/11 attacks, New York Mayor Rudy Guiliani (right) was hailed by many for his leadership during the crisis. To his left is Defense Secretary Donald Rumsfeld. (Photo U.S. Department of Defense)

11.8.4 Personal Example

All of us have a tendency to mimic the behavior of people whom we respect, regardless of their official position. Effective leaders leverage this tendency. All leaders would benefit from becoming more aware of the significant impact their words and actions have on others. Soliciting feedback from others (not during a crisis, but during normal times) is an excellent way to assess how others perceive and respond to the example you set. Leading by personal example in a crisis becomes even more important as workers look to emulate their leaders as their own worlds have become disoriented causing them to question their value system.

11.8.5 Character

Synonymous with integrity, this trait of conscious moral behavior defines who you are when no one is watching. At a minimum, character implies telling the truth, being consistent in word and deed, treating people with dignity, avoiding actions that even hint at impropriety, and exercising self-control in the areas of morality and self-indulgence. Workers will likely tolerate honest mistakes, but may find major moral lapses harder to deal with, especially if they cause substantial harm, i.e., increasing pain and suffering among the affected population. These lapses can have a lasting negative impact on your reputation, erode your ability to lead, and increase the emotional impact of a crisis.

11.8.6 Competence

That leaders should be technically capable of handling their positions is obvious almost to the point that it doesn't need to be mentioned. No amount of personality, political skills, or cracker-barrel wit can disguise or overcome a deficit in basic technical and managerial competence. And almost nothing can multiply worker anxieties and reduce confidence more during a crisis than a leader who is perceived to be marginally competent. Competent leaders instill confidence and remove doubt and fear. Leaders can bolster their strengths and shore up weaknesses by continuously assessing their performance to ensure they have no glaring developmental needs or experiential shortcomings that hinder them from doing their job well in normal circumstances (those hindrances become magnified during a crisis). Developmental activities, such as reading professional books and magazines (everything from biographies to expert discussions of contemporary management subjects), attending continuing education courses and executive training, and taking high-risk, high-payoff assignments are all ways in which leaders can build and maintain competence.

A recent glaring example of this was the removal of the director of FEMA during the early days of the Hurricane Katrina response. Director Michael Brown was politically savvy, articulate, and witty, but none of this could overcome his distinct lack of emergency management and crisis leadership competence. One of his greatest failings was that he failed to lead from the front as he chose not to deploy to New Orleans to take personal charge of the rapidly deteriorating events. This subject of Director Brown's performance continues to be debated to this day as he continues to claim that he was politically sabotaged by both DHS leadership and the Bush Administration White House and there is some documentation to support his claims.

11.8.7 Courage

It takes a high measure of courage to tell the truth under difficult circumstances, to make hard decisions, to answer tough questions, to face the unhappy crowd, and to accept responsibility. But these are the things asked of leaders when a crisis occurs. Those that are impacted by the event will want to know that you will go to bat for them and do the right thing regardless of the consequences or the political environment. Given the leadership failures and scandals among a few corporate giants during the past decade, courage is not always a common trait among senior leaders. A leader without courage courts disaster when a crisis strikes. To build a storehouse of personal courage, leaders can first develop a clear code of personal values, ethics, and standards. They can then relate each situation or decision to this code and follow their conscience to do what is right under the circumstances. This sounds simple, but it is one of the most difficult challenges that many leaders face because of their overriding fear of negative consequences that might affect their reputation, security, and future. There are numerous leaders who took a courageous position during a crisis and then subsequently lost their job as the organization chose to move on after the crisis with new leadership. But, with their honor and integrity intact, most of these leaders easily found new work because their reputations remained intact with the added attraction of being "battled hardened."

11.8.8 Decisiveness

BOX 11.4 IT'S LIKE A WAR ZONE OUT THERE

"The military's single peacetime focus is preparing for combat, the ultimate crisis situation because it involves life and death. A major element of the military's training teaches soldiers how to deal with the range of emotions they will experience before, during, and after combat. These emotions generally include horror, apprehension, grief, rage, revenge, loneliness, sadness, repulsion, vigilance, anguish, and guilt. Military leaders know these emotions will be experienced and must be controlled or the soldiers will not be able to function on the battlefield.

"Combat leaders must learn to deal with their own emotions as well as with the emotions of the soldiers under their charge. This is the same challenge civilian leaders confront during a crisis, and they can expect the same kinds of emotional chaos to flow over the people in their organization and themselves."

Gene Klann
Crisis Leadership (2003)

During a crisis, even a wrong decision that promotes action is better than doing nothing. Influential decision making means gathering information and getting input as soon as possible, knowing and accepting that all the information needed to make the decision isn't available, accepting that there are risks involved, getting recommendations from others, listening to gut feelings, and making a timely decision because it needs to be made.

There are many other skills, traits, and perspectives that leaders can develop and sharpen to enhance their personal influence so that it's more effective during routine operations and during a crisis. Some obvious examples include being adaptable to change, promoting rewards and recognition, maintaining consistency and fairness in discipline, being enthusiastic and optimistic, keeping a sense of humor, endorsing a professional development process for employees, and leveraging strengths that a crisis may demand (such as fluency in a second language or extraordinary computer skills).

11.9 LEADERSHIP'S ROLE IN A CRISIS

Leading in a crisis can be extremely difficult and challenging. Leaders set the tone by their example and conduct. Emergency managers who have led in such circumstances describe the experience as highly developmental—a benchmark in their professional careers. But, what does effective leadership during a crisis look like? There may be as many descriptions of leadership, and crisis leadership, as there are executive coaches and management gurus. In some of its educational activities, CCL has described leadership as a process of influence in which managers interact with workers, contemporaries, and others in the organization in collective pursuit of a common goal.

Ideally, all of us would balance our intellectual, physical, spiritual, and emotional lives all of the time. But that's a difficult job, particularly when a crisis creates an imbalance and tips the scale toward the emotional end. This creates a special challenge for emergency managers who must provide leadership to those who are in a state of emotional turmoil. Occupying a designated leadership position isn't the same thing as being a leader, doesn't provide leadership on its own, and doesn't prove that the person in that position has the skills, knowledge, and ability to be an effective leader. We can all remember FEMA Director Brown in the days following Hurricane Katrina. As the director of FEMA, he held an important leadership position, but failed to demonstrate that he was the leader that was needed at the moment of crisis, losing the confidence of those both above and below him, and subsequently losing his job.

Leaders who view themselves as successful because of position, salary, or longevity, but leave a high body count of former employees bobbing in their wake are often surprised to find their careers derailed or sidelined. Nothing separates such leaders from their illusions as quickly and sharply as a crisis because it's then they realize they haven't built the skills necessary to lead effectively during such traumatic incidents.

An organization's senior leadership is a key factor before, during, and after a crisis, and its quality can determine the length, severity, and ultimate consequences of the crisis. Leaders set the tone by their example and conduct during the crisis situation. By paying attention to the components of influence (especially communication, clarity of vision and values, and caring), leaders can have a significant positive impact on the very human, emotionally charged climate that accompanies a crisis. That in turn can reduce the negative impact and duration of a crisis for the benefit of the organization.

11.9.1 Emotional Intelligence

In the early 1990s, Peter Salovey and John Mayer (1990) introduced a concept that has since become known as *emotional intelligence*, or EQ. This kind of intelligence describes a person's ability to understand his or her own emotions as well as the emotions of others, and it describes a person's ability to act and behave according to that understanding. Several books popularized emotional intelligence, including Goleman (1995), who listed social awareness and social skills as two of the five competency areas of EQ. Social awareness includes empathy, political savvy, appreciating diverse perspectives, and being sensitive to other people's feelings. It's essentially the ability to detect and be responsive to the emotions, moods, intentions, needs, and desires of others. Social skills include being discerning about what is going on around you, communicating well-timed comments and observations, possessing a general competence level at a variety of social settings, and positively managing emotions in others. CCL research has found that one cause of derailment involves deficits in interpersonal relationships, or what might be described as emotional intelligence. Alternatively, effective leaders are similar in that they all possess a high degree of EQ.

11.9.2 Competence

BOX 11.5 QUOTE

"No amount of personality, political skills, or cracker-barrel wit can disguise or overcome a deficit in basic technical and managerial competence. And almost nothing can multiply employee anxieties and reduce confidence more during crisis than a leader who is perceived to be marginally competent."

Gene Klann
Crisis Leadership (2003)

In section 11.8.6, we extensively discussed the need for a leader to possess competence. As discussed earlier, Michael Brown, the former FEMA director, was accused not only of failing to lead, but also for lacking the technical competency to be director of FEMA. Following Hurricane Katrina, many complained that Director Brown's qualifications were limited to running a racing horse association. Worth noting here is an interesting paradox; some of the loudest complaints came from members of Congress where the Senate confirmed Brown as director of FEMA. The confirmation process failed to recognize that Brown lacked both the leadership and competency to be an effective director of FEMA in a catastrophe. Competent leaders instill confidence and remove doubt and fear. In ICS training, on-scene commanders are taught the concept of *train, test and trust*. Train your people, test them to make sure they know how to do a job, and then trust them to do it.

11.9.3 Courage

Previously discussed in section 11.8.7, it takes a high measure of courage to tell the truth under difficult circumstances, to make hard decisions, to answer tough questions, to face the unhappy crowd, and to accept responsibility.

11.9.4 Decisiveness

As previously discussed in section 11.8.8, even a wrong decision that promotes action is better than doing nothing. Influential decision making means gathering information and getting input as soon as possible, knowing that all the information needed to make the decision isn't available, accepting that there are risks involved, getting recommendations from others, listening to gut feelings, and making the decisions that need to be made.

11.10 PREPARING FOR CRISIS

Communication, clear vision and values, and the fostering of effective, caring relationships all directly affect how people will emotionally respond to a crisis. These three capabilities strengthen a leader's ability to help the organization and its people to weather a crisis. But more must be done. Effective leaders prepare their organizations for crisis.

Traditionally, books and articles about crisis management place a great deal of emphasis on management actions to be taken in preparation for a crisis. This focus on management functions implies that you can prepare for a crisis by writing a plan and then executing it when the crisis occurs. Certainly, an effective leader is competent in such functions as planning, organizing, staffing, budgeting, controlling, and directing. But a narrow emphasis on management strategy and planning ignores the leadership necessary for putting the plan into action. That kind of approach sidesteps the human element that plays such a large role during a crisis: the needs, emotions, and behaviors of people at all levels of the organization.

11.11 LEADING DURING A CRISIS

The time for planning is over. The crisis has struck and can be felt throughout the organization. It doesn't matter whether or not you have developed a crisis action plan (as described in Klann (2003), Appendix A, pp. 75–80). A crisis sets its own timetable. It doesn't wait for a plan to be drawn up or for leadership skills to be developed. But, if an organization has developed a strategy for dealing with a crisis, and if much of the leadership efforts focused on the human elements (having assessed and developed communication and relationship skills, developed a clarity of vision that's aligned with personal values and the values of your organization, and worked to foster and maintain effective relationships), then there is an opportunity to successfully lead through a volatile situation.

11.12 RECOVERY AND REBUILDING

After the emergency phase of a catastrophe has passed, a community will move into the restoration and reconstruction phases. While we may have moved past the phase of a catastrophe that represents an imminent threat to life, strong leadership skills will still be needed. Remember, we discussed the Recovery Model in Chapter 9 where we looked at Haas et al.'s Model of the Disaster/Catastrophe Recovery Process (1977) (Figure 11.5). Notice that in a disaster, the model estimates that Phase II of Reconstruction can extend out about 10 years after the incident. In a catastrophe, this can take a generation, i.e., 20 years or more. Clearly, recovering and rebuilding after a catastrophe is going to be a marathon, not a sprint.

Continuous assessment of progress is one method of demonstrating leadership during the recovery and rebuilding phase of a catastrophe. This helps to ensure that the process is progressing in a manner consistent with the community leadership's wishes. Part of the continuous assessment is to conduct after-action reviews or "lessons learned" events. When conducting these events, it is vital to involve all stakeholders so no group feels marginalized. Now, more than during the emergency phase,

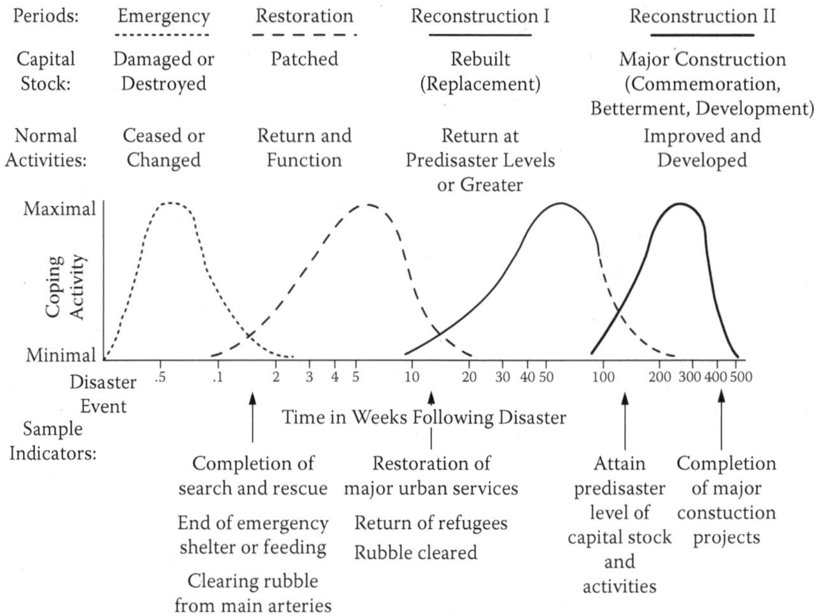

FIGURE 11.5 A model of the Disaster Recovery Process. (From Haas et al. 1977. *Reconstruction following disaster.* Cambridge, MA: MIT Press. With permission.)

an effective leader will need to enlist of the support of others. Applying the three "Cs" model (communication, clarity of vision and values, and caring) is an effective method of accomplishing this.

11.13 WHAT LEADERS CAN DO TO TAKE CARE OF THEMSELVES DURING A CRISIS

During a crisis, leaders are often concerned with the emotional turmoil of their direct reports and others in the organization. That's the way it should be, as the human impact of a crisis has a good deal to do with whether or not an organization can seize opportunity from the turbulence it finds itself in, or if it can even survive the crisis at all. Leaders are also absorbed with managerial tasks designed to keep the organization running on a functional level. Both of these leadership tasks are important, but it's equally important for leaders to take care of themselves physically and psychologically during a crisis. They should be aware of their own emotional turmoil, its effect on their behavior, and its influence on their leadership abilities. Appendix B in Klann's *Crisis*

Leadership (2003) provides a list of ideas, practices, and reminders that can help leaders keep the perspective they need to bring their people and their organization through a crisis.

11.14 DISCUSSION QUESTIONS

1. Do you agree with Lagadec that the nature of catastrophes is changing?
2. Do you think Rapid Reflection Forces would be effective?
3. Are the federal emergency management framework and the crisis leadership skills described in this chapter compatible in a catastrophe?
4. What makes a leader?
5. To define what makes a leader, consider traits of leadership, their relationship with the concept of emotional intelligence, and answer the following:
    - Write down 10 traits of a successful leader.
    - How many of these 10 traits do you have?
    - How many of these traits relate to emotional intelligence?
    - What do you have to do to demonstrate more of these traits?

REFERENCES

Braden, V., J. Cooper II, M. Klingele, J. Powell, and M. Robbins. 2005. *Crisis: A leadership opportunity*. Cambridge, MA: Harvard University Press. Online at: https://www.afresearch.org/skins/rims/q_mod_be0e99f3-fc56-4ccb-8dfe-670c0822a153/q_act_downloadpaper/q_obj_26bb88b3-1bf8-4bf3-8a75-0bcebf14e04d/display.aspx?rs=enginespage (accessed January 10, 2010).

Goleman, D. 1995. *Emotional intelligence*. New York: Bantam Books.

Haas, J.E., R.W. Kates, and M.J. Bowden (eds.). 1977. *Reconstruction following disaster*. Cambridge, MA: MIT Press.

Klann, G. 2003. *Crisis leadership: Using military lessons, organizational experiences, and the power of influence to lessen the impact of chaos on the people you lead*. Greensboro, NC: CCL Press.

Lagadec, E. 2007. *Unconventional crisis, unconventional responses: Reforming leadership in the age of catastrophic crisis and hypercomplexity*. Washington, D.C.: Center for Transatlantic Relations.

Salovey, P. and J. Mayer. 1990. Emotional intelligence. *Journal of Imagination, Cognition, and Personality*. 9: 185–211.

Yukl, G. 1989. *Leadership in organizations*. Englewood Cliffs, NJ: Prentice Hall.

V Conclusion

Here are some closing thoughts on being an effective crisis leader:

1. **Leaders must be engaged before, during, and after a crisis.** If the reader thinks they are going to be able to put on their "crisis leader" hat when something really bad happens, it isn't going to go well. Effective crisis leaders hone their leadership engagement skills every day in all kinds of situations. One must be a respected leader before the crisis if people are expected to follow one's leadership in a crisis.
2. **Leaders must be familiar with federal emergency management framework (NRF, NIMS, and ICS).** For better or worse, the federal response system is probably going to be around when the next crisis befalls your community. While we can debate how relevant federal response systems will be in the next catastrophe the United States will face, emergency managers are going to attempt to utilize it, so one needs to understand how their crisis leadership is and is not consistent with federal emergency management framework and be prepared to explain variances from this and why it is necessary to do so.
3. **Crisis leaders must be skilled in communication, clarifying vision and values, and demonstrate caring at all times, not just during a crisis.** Leaders can't be a chameleon, where one fails to demonstrate the three "Cs" (communication, clarity of vision and values, and caring relationships) in how they conduct themselves in the regular course of business and then expect people to accept their leadership for demonstrating the three "Cs" during a crisis, as people will question their commitment.
4. **Effective crisis leaders always have a plan.** If there is any hope for being an effective crisis leader, one must invest in developing a Crisis Action Plan, one tailored to play to your strengths and to minimize your weaknesses.
5. **You are no good to anyone if you don't take care of yourself in a crisis.** Leaders must develop a self-awareness of their own emotional turmoil, its effect on their behavior, its influence on their

leadership abilities, and then take effective countermeasures to maintain their effectiveness.

6. **Leaders need to take time to reflect on their effectiveness.** The toughest critic of one's leadership skills must be oneself. Leaders must constantly strive to become a more effective leader as only the best leaders rise to be outstanding crisis leaders.

Epilogue

Through the five sections of this book, the reader has been introduced to the many political, legal, economic, ethical, programmatic, and other issues that influence how we plan for and respond to a catastrophe.

- **Section I**: Because it is important to understand the unique characteristics of catastrophes, in Section I, we reviewed the history of catastrophic events both in and outside the United States, how catastrophes differ from disasters and emergencies, and how they are all part of the emergency management continuum. We also explored the varying definitions of catastrophes and their political and societal implications.
- **Section II**: Here we introduced the reader to the inherent ethical, political, and legal issues associated with catastrophic events. With respect to ethics, this part of the book addressed the main ethical and value dilemmas and quandaries that will likely be faced before, during, and after a catastrophe. We also explored the legal framework associated with government response to catastrophes that includes the use of the military for domestic response, suspension of civil rights, and federal control of industrial output. Political factors, as well as organizational dynamics, were included to provide a basis for understanding the complex environment in which preparing for extreme events, such as a catastrophe, may take place. The literature, reflected in political science and public administration, provided us with insights into conflicts that arise in highly stressful events and the nature of the problems that evolve from our attempts to deal with a catastrophe.
- **Section III**: In this section, we explored the postcatastrophic environment that an emergency manager might well experience. We learned that many, if not all, environs will be altered by the event. Government, industrial, health and safety processes, and critical infrastructure/key resources (CI/KR) systems will cease to operate as intended for extended periods of time. We investigated the impact a catastrophe might have on logistics, CI/KR, and mass care.
- **Section IV**: While we learned in previous parts of the book that a catastrophe will certainly exceed available response, recovery,

and reconstruction capabilities, here we explored planning strategies and skills an emergency manager can employ to mitigate the effects of such an extreme event. Issues that were discussed included the use of volunteers and unconventional sources of assistance, implications on mass care issues, and an understanding of the political challenges associated with response, recovery, and reconstruction.

- **Section V:** We learned that faced with a catastrophe, an emergency manager will need to utilize crisis leadership skills that he/she may not have had to use before. We learned how to lead and influence others in a catastrophic situation by increasing your crisis leadership skills and abilities, also known as our *emotional intelligence*.

Now, at the conclusion of this book, the reader should have a solid understanding of principles of catastrophic planning and response, including its historical development, elements, structures, and functions. This book provides a foundation for the continued study of and critical reflection on catastrophic planning and response. The future of catastrophic planning and response is being shaped by the actions we take today as citizens, academicians, professionals, and students. It is quite possible that catastrophic planning will change substantially in the coming years, especially if the United States has to face another Hurricane Katrina or worse incident.

A strong grasp of the fundamentals and objective apolitical analysis of how catastrophic planning has developed in recent years is essential to a successful future as a professional in the field of emergency management. Hopefully, the reader will learn from the history of catastrophic planning and recognize that hard lessons do not have to be learned again. The continuing incident response issues evidenced by the response to Hurricane Katrina show that catastrophic planning continues to be a work in progress. In fact, the study of catastrophic planning and response is probably just in its infancy. What is most needed now are emergency management professionals who understand the catastrophic environment and who can make tough decisions and changes where necessary for a safe and secure future.

Index